Systems Biology

Topical Conference at the 2008 AIChE Annual Meeting

Philadelphia, Pennsylvania
16 - 21 November 2008

ISBN: 978-1-61567-207-3

Printed from e-media with permission by:

Curran Associates, Inc.
57 Morehouse Lane
Red Hook, NY 12571
www.proceedings.com

Some format issues inherent in the e-media version may also appear in this print version.

Printed by Curran Associates, Inc. (2009)

AIChE
3 Park Avenue
New York, NY 10016-5991

Phone: (203) 702-7660
Fax: (203) 775-5177

www.aiche.org

TABLE OF CONTENTS

51a Do We Know Enough to Reconstruct Life In Silico?

Michael L. Shuler, Chemical and Biomolecular Engineering and Biomedical Engineering, Cornell University, 120 Olin Hall, Ithaca, NY 14853-5201, Jordan C. Atlas, School of Chemical and Biomolecular Engineering, Cornell University, 120 Olin Hall, Ithaca, NY 14853, and Evgeni V. Nikolaev, Department of Pathology, Anatomy, and Cell Biology, Thomas Jefferson University, 1020 Locust Street, Philadelphia, PA 19107.

The question of "What is essential for life?" is one of the most fundamental questions humankind faces. If we fully understand a system, we should be able to reconstruct it. We propose to construct in silico a model of a minimal chemoheterotrophic bacterial cell. Such a model is key to fully understanding and identifying underlying regulatory and organizational concepts central to life. The success of whole organism genome sequencing and high-throughput measurements provides an opportunity for system-level analysis of whole organisms, or what has been termed "systems biology". Systems biology investigates the behavior of all of the elements in a biological system while it is functioning. As a systems biology approach, the Minimal Cell Model (MCM) depicts the total functionality of a minimal cell and its explicit response to perturbations in its environment. In this paper we describe our approach to creating such a model and results demonstrating feasibility. We also discuss constraints on identifying all essential functions.

I. Purpose

We propose a dynamic modeling framework to integrate genomic detail and cellular physiology within functionally complete 'hybrid' bacterial cell models. An initial step in this approach is the development of a whole-cell coarse-grained model which explicitly links DNA replication, metabolism, and cell geometry with the external environment. A hybrid model can then be constructed from chemically-detailed and genome-specific subsystems, called modules, inserted into the original coarse-grained model. We have demonstrated this possibility with modules for nucleotides [1] and lipids [2]. We use the sensitivity analysis of the original coarse-grained model to identify which pseudo-molecular processes should be de-lumped into molecularly detailed mathematical modules to implement a particular biological function.

A minimal cell is a hypothetical entity defined by the essential functions required for life [1]. Although others have the goal of experimentally constructing a minimal cell [3], we seek to identify a minimum number of genes necessary and sufficient for the cell to divide and grow continuously in a rich environment with preformed nutrients and constant temperature and pH. The model, which contains kinetic, thermodynamic, and stoichiometric constraints, is used as a tool to identify the organizing principles which relate the dynamic non-linear functioning of the cell to the genome sequence. We recognize that a unique set of minimal genes may not exist, but we believe a unique set of essential functions exist and that we can generate a candidate minimal gene set to accomplish these functions. Functions may be chemical or physical such as constraints on DNA packing.

II. Approach

The project proposed here includes three main parts: 1) Development of novel algorithms for stability and sensitivity analysis of hybrid cell models, 2) Implementation of a system for parameter estimation, and 3) Construction of a genomically and chemically detailed Minimal Cell Model.

III. Results

Using an existing coarse-grained model of E. coli [4] as a basis, we have developed a framework for analyzing the stability of hybrid cell models [5] as well as a technique for sensitivity analysis in this class of models using an extension of classical Metabolic Control Analysis [6]. Currently, we are extending a statistical mechanics method for parameter estimation [7] to our models.

References

[1] Castellanos, M., D.B. Wilson, M.L. Shuler. (2004) A modular minimal cell model: purine and pyrimidine transport and metabolism. PNAS 101:6681-6686

[2] Castellanos, M., K. Kushiro, S.K. Lai, and M.L. Shuler. 2007. A genomically/chemically complete module for synthesis of lipid membrane in a minimal cell. Biotechnol. Bioeng. 97:397-409

[3] Forster, A.C., G.M. Church. (2006) Towards Synthesis of a Minimal Cell. Molecular Systems Biology 2, 45.

[4] Shuler, M.L. (2005) Computer Models of Bacterial Cells to Integrate Genomic Detail with Cell Physiology. Proceedings of the KBM International Symposium on Microorganisms and Human Well-Being, Seoul Korea.

[5] Nikolaev, E.V., J.C. Atlas, M.L. Shuler. (2006) Computer Models of Bacterial Cells: From Generalized Coarse-Grained to Genome-Specific Modular Models. Journal of Physical Conference Series 46, 322-326.

[6] Nikolaev, E.V., J.C. Atlas, M.L. Shuler. (2007) Sensitivity and control analysis of periodically forced reaction networks using the Green's function method. J. Theoretical Biol. 247:442-461

[7] Brown, K.D., J.P. Sethna. (2003) Statistical Mechanical Approaches to Models with Many Poorly Known Parameters. Phys. Rev. E.68

Acknowledgements: This work was funded by DOE grant DE-FG02-04ER63806 and by NYSTAR, the New York State Office of Science, Technology, and Academic Research. Jordan Atlas gratefully acknowledges support from the DOE Computational Science Graduate Fellowship Program of the

Office of Science and National Nuclear Security Administration in the DOE under contract DE-FG02-97ER25308.

1School of Chemical and Biomolecular Engineering, Cornell University, Ithaca, NY 14850

2Department of Biomedical Engineering, Cornell University, Ithaca, NY 14850

51b Using Computations to Reconstruct, Analyze and Redesign Metabolic Systems

Costas Maranas, Chemical Engineering, Penn State, 112 Fenske Lab, University Park, PA 16802

Metabolism is defined as the full complement of chemical transformations in living systems. Systems biology techniques are increasingly being used to elucidate and quantify the full range of molecules (e. g., metabolite concentrations) and transformations (e.g., reaction fluxes) at play. In this talk we will explore how computations can shed light onto the interdependence, fragility and redundancy of metabolism in living systems. We will discuss how we can pinpoint the essential core of genes needed to ensure life's processes to better understand the organizational principles of metabolism and provide insights into the design of minimal organisms. We will next explore how we can speed up the process of building organism-specific metabolic models by automatically filling in connectivity gaps and restoring consistency with gene essentiality experiments. With the metabolic reconstructions in place, we will next describe how we can use computations to decipher metabolic flows given NMR or GC/MS spectra derived from ^{13}C labeling experiments. Optimization based techniques will finally be illustrated for strain optimization case-studies by classifying all fluxes in a metabolic model depending upon whether or not they must increase, decrease, or become equal to zero to meet a pre-specified overproduction target.

Web Page: maranas.che.psu.edu/

51c Finding Needles In Haystacks: Functional Genomics and Extremophile Biotechnology

Robert M. Kelly, Chemical and Biomolecular Engineering, North Carolina State University, 911 Partners Way, EB-1, Box 7905, North Carolina State University, Raleigh, NC 27695-7905

Current rates for sequencing microbial genomes are far outpacing the capacity to understand the corresponding biological complexity encoded in the DNA. Much attention is focused on in silico methods for extracting insights into biological function. However, it is becoming clear that "low throughput" efforts are needed to complement the "high throughput" capabilities associated with systems biology. In fact, a critical challenge now is how to balance modern and classical approaches to obtain new insights into the relationship between genotype and phenotype.

In our laboratory, we use functional genomics approaches to study extremely thermophilic microorganisms. Our model microorganisms include archaea, such as *Pyrococcus furiosus* (T_{opt} 98-100° C), *Metallosphaera sedula* (T_{opt} 73°C, pH_{opt} 2.0), and *Sulfolobus solfataricus* (T_{opt} 80°C, pH_{opt} 3.0) and bacteria, such as *Thermotoga maritima* (T_{opt} 80°C) and *Caldicellulosiruptor saccharolyticus* (T_{opt} 70° C). These microorganisms have significant biotechnological potential as sources of thermostable biocatalysts, as biohydrogen producers, for recovery of base and precious metals, and for unique metabolic characteristics. Using genome sequence information, transcriptional response analysis, biochemical characterization of specific proteins, and microbiological assays, we have uncovered numerous leads with biotechnological relevance. In addition, this has enabled us to provide definitive annotation of many ORFs that otherwise would remain as "hypothetical" or "conserved hypothetical" proteins in our model systems. Discussed here is how modern and classical methods are being used to study extreme thermophiles, an approach that can be extended to less extremophilic microorganisms and cells. Examples provided will include biocatalyst discovery, metagenomics, interspecies interactions, metabolic pathway identification, secretome analysis, and comparative bioenergetics.

101a Predicting the Evolutionary Landscape of Multicellular Phenotypes

Anand R. Asthagiri, Chemical Engineering, California Institute of Technology, 1200 E. California Blvd MC 210-41, Pasadena, CA 91125

Networks of biological signals guide cells to form multicellular patterns and structures. Understanding the design and function of these complex networks is a fundamental challenge in developmental biology and has clear implications for biomedical applications, such as tissue engineering and regenerative medicine. Signaling networks are composed of highly interconnected pathways involving numerous molecular components. Precisely to what extent multicellular structures are susceptible to quantitative variations in underlying signals and to what extent Nature has utilized this mechanism of "quantitative diversification" during evolution are unclear. I will describe a computational framework that we have developed to explore and to quantify the multicellular diversity that emerges from signaling perturbations. We have applied this method to study vulval development in C. elegans. The approach is not only effective in predicting the molecular genetics of multicellular patterning, but also gauges the capacity of this signaling network for creating phenotypic diversity. In fact, model predictions strongly correlate to multicellular phenotypes observed across ten species related to C. elegans. These results suggest that systems-level modeling can shed insight into the evolutionary trajectories of regulatory networks that gave rise to divergent multicellular phenotypes.

101b Multiscale Stochastic Dynamical Models Calibrated from AFM Single-Molecule Time Series and Computer Simulations

Christopher P. Calderon[1], Ching-Hwa Kiang[2], Dennis D. Cox[3], Nolan C. Harris[2], Wei-Hung Chen[2], and Kuan-Jiuh Lin[4]. (1) Computational and Applied Mathematics, Rice University, 6100 Main St, Houston, TX 77251-1892, (2) Physics & Astronomy, Rice University, 6100 Main St, Houston, TX 77251-1892, (3) Statistics, Rice University, 6100 Main St, Houston, TX 77251-1892, (4) Chemistry, National Chung-Hsing University

The stochastic dynamics associated with stretching both single-stranded poly(dA) DNA and synthetic titin molecules (each molecule contained 8 identical serially linked I27 domains) were approximated using maximum likelihood state-space techniques. The models were fit using experimental atomic force microscopy (AFM) time series. The measurement noise associated with the AFM apparatus was quantified using time domain techniques which exploited a time scale separation that exists in the data [1,2]. All results reported in this study analyzed repeat experiments on the same underlying molecule, i. e. once a single-molecule is captured by the AFM tip, it is retained and a series of experiments are conducted with the same molecule. The fitted stochastic models are fairly flexible, the major assumptions made about the dynamics were that overdamped diffusion models can be used to adequately approximate the true process and that the drift and diffusion coefficient (whose global functional form is assumed unknown) of these models are well-behaved, e.g. twice differentiable. The assumptions are tested in an a posteriori fashion using goodness-of-fit tests given the data and the models estimated from the observed data (the tests employed check for non-Markovian effects). The unfolding of the I27 domain of titin was also investigated [2] using steered molecule dynamics simulations in order to test a variety of hypothesis suggested by the experiments.

The estimated pathwise dynamical models can be used for a variety of purposes. They can be used to simulate and quantify how solvent induced thermal noise would affect the distribution of certain observables like the nonequilibrium work associated with stretching the captured molecule. This is relevant in situations where collecting a large number of force extension curves (FECs) is difficult due to experimental complications (e.g., in some systems retaining the same molecule for a large number of repeated force extension cycles is difficult).

The models can also be used to indirectly determine when certain important structural transitions occur and also quantify how conformational degrees of freedom (which typically are associated with slower time scales) influence the multiscale stochastic dynamics [1,2,3]. This indirect inference requires one to collect a batch of FECs and compare the "rules" governing the resulting models. In some cases, the measured drift contains a signature of a transition and in others a sudden thermal noise magnitude change acts as a fingerprint of a structural change. The primary appeal of this type of work is that the methods developed more fully utilize the wealth of information contained in the time series collected from single-molecule force spectroscopy and computer simulations. The diffusion model functions facilitate physical understanding of the process dynamics and aid in interpreting the experiment and simulation results. Quantitative information about stochastic dynamical responses of single-molecules induced by different stimuli (mechanical and/or chemical) can possibly enhance targeted drug design which exploit recent rapid advances in nanotechnology.

[1] C.P. Calderon, W.-H. Chen, K-.J. Lin , and C.-H. Kiang (submitted).

[2] C.P. Calderon, N.C. Harris, C.-H. Kiang and D. D. Cox (in preparation).

[3] C.P. Calderon and R. Chelli. J. Chem. Phys., 128:145103 (2008).

Web Page: www.stat.rice.edu/~cpc1/

101c Ligand Specificity of C-Fos Expression Emerges from Spatiotemporal Control of Erbb Network Dynamics

Takashi Nakakuki[1], **Marc R. Birtwistle**[2], Yuko Saeki[1], Noriko Yumoto[1], Nagashima Takeshi[1], Babatunde A. Ogunnaike[2], Hatakeyama Mariko[1], and Boris N. Kholodenko[3]. (1) Advanced Computational Sciences, RIKEN Advanced Science Institute, Yokohama, Japan, (2) Chemical Engineering, University of Delaware, 150 Academy St., Newark, DE 19716, (3) Pathology, Anatomy, and Cell Biology, Thomas Jefferson University, 1020 Locust St., Philadelphia, PA 19107

Even though ligand-induced activation of different ErbB receptors leads to distinct cellular outcomes, there is as yet no mechanistic understanding of how receptor-specific responses emerge when signals propagate through the same core components. The current work focuses on this topic in the context of how two prototypical ErbB receptor ligands, epidermal growth factor (EGF) and heregulin (HRG), induce differential expression of the transcription factor c-Fos through the core ERK pathway in MCF-7 breast cancer cells. Although EGF and HRG induce transient versus sustained ERK activation, respectively, we found that the duration of *c-fos* mRNA expression is surprisingly similar for both ligands, whereas c-Fos protein level displayed all (HRG) or none (EGF) behavior. In this presentation, we will discuss our mechanistic modeling which suggests, and our immunofluorescence experiments which confirm, that induction of the nuclear phosphatase *mkp1* causes nuclear ERK activation to be transient for both ligands. Since active, nuclear ERK stimulates *c-fos* transcription, the transience of active, nuclear ERK for both ligands explains the similar *c-fos* mRNA expression durations. We will also show how combinatorial siRNA knockdown of *mkp* family proteins verifies the key role these phosphatases play in controlling the *c-fos* mRNA expression dynamics. Since de novo c-Fos protein is stabilized by active ERK, we deduce that the large, HRG-induced c-Fos response arises from sustained active ERK in the cytoplasm. Thus, we conclude that the all-or-none c-Fos responses arise from ligand-dependent spatiotemporal control of ERK activity. Longer time course gene chip data suggest that these quantitative differences in expression of immediate early transcription factors such as *c-fos* leads to qualitative differences in subsequent waves of gene expression, and consequently differential cell fate decisions.

101d Synthesis of Metabolic Pathways Using Thermodynamic Constraints and Parametric Sensitivity Analysis

Frank P. Marealle[1], Ye Wang[1], **Jorge E. Gatica**[1], and María Rosa Hernández[2]. (1) Chemical and Biomedical Engineering, Cleveland State University, 455 Stilwell Hall, 2121 Euclid Avenue, Cleveland, OH 44115-2214, (2) Chemistry and Chemical Engineering, Universidad Nacional de Tucuman, Tucuman, 4000, Argentina

In order to understand quantitatively how metabolic processes work, a mechanistic model is often formulated through biochemical mass balances and reaction kinetics. The modeling of a recognized, although not fully characterized yet, metabolic pathway is selected as an illustrative case study for this paper: the astrocyte-neuron lactate shuttle in mammalian brain metabolism. Extension to other biomedical and biotechnology applications is discussed.

The brain is the supervisory center of the nervous system in all mammals. Regulation of brain metabolism and cerebral blood flow involves complex control systems with many several interacting variables at both cellular and organ levels. Quantitative understanding of the spatially and temporally heterogeneous brain control mechanisms during internal and external stimuli requires the development and validation of a computational (mathematical) model of the metabolic processes in the brain tissue.

Unlike metabolic pathways in liver, heart and skeletal-muscle tissues, metabolism in mammalian brain tissue is widely believed to follow a compartmental model. Indeed, the conventional view that glucose oxidation fuels most activity-associated energy metabolism in neurons is challenged by the astrocyte-neuron lactate shuttle (ANLSH) hypothesis [1]. According to the ANLSH, neural activity increases the extra cellular concentration of glutamate, whose uptake by glia stimulates Na+ - K+ ATPase and glutamate synthetase activity. This, in turn, stimulates glial glycolysis leading to the lactate production in the cytoplasm of astrocytes. Lactate accumulates in the intercellular space and it is eventually transported into the neuronal cytoplasm. This hypothesis suggests that astrocyte-produced lactate is the main oxidative substrate for neurons, particularly during neuronal activation. The formulation of a mechanistic model of the astrocyte-neuron lactate shuttle is analyzed in this paper. The model structure consists of neurons, astrocytes, and a surrounding capillary network.

The resulting dynamic model is a highly nonlinear distributed parameter system with model equations that combine several chemical species, reactions and biochemical fluxes. Due to the difficulty in physically obtaining reaction rate constants for individual biochemical reactions, flux balances are often employed to reduce the number of unknown parameters [2]. Although flux balances can be valuable predictive tools, their use with the inclusion of proper thermodynamic constraints in pathway analysis can be used to more reliably estimate biochemical reaction rate parameters, and yield a robust dynamic model [3]. The integration of these principles within the formulation and verification of a metabolic model is demonstrated in this work.

References

[1] Magistretti, P.J., L.Pellerin, D.L. Rothman, and R.G. Shulman "Energy on demand," Science 283:496497 (1999).

[2] Patel, A.B., R.A. de Graaf, G.F. Mason, T. Kanamatsu, D.L. Rothman, R.G. Shulman, and K.L. Behar "Glutamatergic neurotransmission and neuronal glucose oxidation are coupled during intense neuronal activation," J Cereb Blood Flow Metab. 24: 972-985 (2004).

[3] Beard, D.A. "A biophysical model of the mitochondrial respiratory system and oxidative phosphorylation," PLoS Comput Biol 1(4): e36 (2005).

101e Host and Viral Factors Control Stochastic Behavior In HIV-1 Gene Regulation and Latency

Kathryn Miller-Jensen[1], Ron Skupsky[2], John C. Burnett[1], Adam P. Arkin[3], and David V Schaffer[1]. (1) Chemical Engineering, University of California at Berkeley, Berkeley, CA 94720, (2) California Institute for Quantitative Biomedical Research, University of California at Berkeley, Berkeley, CA 94720, (3) Bioengineering, University of California at Berkeley, Berkeley, CA 94720-1462

Upon infection by HIV, a host cell proceeds through a critical choice between replication and latency. Viral replication causes host cell death, while latent infections can persist for decades and represent the most significant obstacle to eliminating HIV from a patient. This fate decision depends on both the chromatin environment at the viral integration site, and the availability of transcription factors within the host cell. We have previously shown that, for some integration positions, HIV promoter activity can result in a bifurcating expression pattern in an infected clonal population, where some cells have high promoter activity while others have low activity. Here we explore how the host transcription factor NF-κB and the viral transactivator of transcription Tat affect expression from the HIV promoter at fixed integration positions. Computational modeling predicts that clonal populations will transition between specific expression profiles if the concentration of either of these factors is systematically varied. We experimentally test our prediction by inducibly expressing Tat or NF-κB in our model system, and directly demonstrate that a clonal population with initially low promoter activity can transition to a bifurcating state. We further show that at this new transcription factor steady-state, stochastic fluctuations drive switching between low and high promoter activity. By quantitatively understanding how transcription factors affect the HIV fate decision, we hope to improve rational design of therapeutic strategies to counteract viral latency.

101f Stochastic Spatial and Non-Spatial Simulations of the Erbb Signaling Pathways Demonstrate the Importance of Spatial Organization In Signal Transduction

Michelle N. Costa[1], Dion Vlachos[2], and Jeremy S. Edwards[1]. (1) Chemical and Nuclear Engineering, University of New Mexico, Albuquerque, NM 87110, (2) Department of Chemical Engineering, University of Delaware, 150 Academy Street, Newark, DE 19716

Overexpression of the ErbB family of receptors leads to an activation of intracellular signaling pathways that are responsible for controlling cellular proliferation, growth, differentiation, and apoptosis. Therefore, overexpression of the ErbB receptors is often associated with cancer, and extensive efforts have been made to understand the signaling events associated with ErbB activation. In recent years, several different research groups have used a systems biology perspective to study ErbB signaling. Although the systems biology computational models have grown in complexity, little work has been done to uncover the spatial-temporal dynamics of receptor interactions and how it affects downstream signaling within this pathway. Herein, we have developed an algorithm which couples a spatial stochastic model of receptor reaction and diffusion on the cell membrane with a nonspatial stochastic model of intercellular reactions and interactions in the cytosol. This framework allowed us to elucidate the effects of downstream signaling due to receptor clustering and spatial heterogeneity on the membrane.

101g Reverse Engineering the Human Platelet

Jeremy E. Purvis[1], Manash S. Chatterjee[2], Lawrence F. Brass[1], and Scott L. Diamond[2]. (1) Medicine, University of Pennsylvania, 5140 Vagelos Research Laboratories, 3340 Smith Walk, Philadelphia, PA 19104, (2) Chemical and Biomolecular Engineering, University of Pennsylvania, 5140 Vagelos Research Laboratories, 3340 Smith Walk, Philadelphia, PA 19104

Computational modeling has enabled systematic analysis of complex biological signaling networks and can lead to the generation of testable hypotheses. Here, we report for the first time the integration of kinetic data, electrochemical calculations, multiple cellular compartments, population versus single-cell behavior, and modular organization of cell function to construct a high-resolution model of platelet homeostasis and activation signaling. The model has distinct modules for calcium release and uptake, phosphoinositide (PI) metabolism, $P2Y_1$ G-protein signaling, and protein kinase C (PKC) regulation of phospholipase C-&[beta] (PLC-&[beta]). The full deterministic model accurately predicted (*i*) steady-state concentrations for intracellular calcium ($Ca^{2+}{}_i$), inositol 1,4,5-trisphosphate (IP_3), diacylglycerol (DAG), phosphatidic acid (PA), phosphatidylinositol (PtdIns), phosphatidylinositol phosphate (PIP), and phosphatidylinositol 4,5-bisphosphate (PIP_2), (*ii*) transient increases in $Ca^{2+}{}_i$, IP_3, and Gq&[alpha]&[middot]GTP in response to ADP, and (*iii*) the volume of the platelet dense tubular system (DTS). A more stringent test of the model involved stochastic simulation of individual platelets which display an asynchronous calcium spiking behavior in response to ADP. The model reproduced the measured frequency distribution of spiking events and demonstrated that the asynchronous spiking was in part a consequence of stochastic fluctuations due to the small volume of the platelet. Given the role of platelets in mediating thrombosis and hemostasis, as well as their contribution to systemic disorders such as inflammation and cancer, the model provides a quantitative framework for predicting platelet behavior and designing patient-specific therapies.

Web Page: bloodsystemsbiology.org

251a Computational Design of Orthogonal Ribosomes In Bacterial Hosts

Lon Chubiz, Chemical and Biomolecular Engineering, University of Illinois, 600 South Matthews Ave., 211 Roger Adams Laboratory, Box C-3, Urbana, IL 61801 and Christopher V. Rao, Department of Chemical Engineering, University of Illinois, 600 South Matthews Ave., 211 Roger Adams Laboratory, Box C-3, Urbana, IL 61801.

Gene expression involves two steps, transcription and translation. While a number of genetic tools exist for reprogramming transcription in cells, far fewer tools exist for translation. Of the tools available in bacteria, the most popular are riboregulators, both cis and trans-activating, and orthogonal ribosomes (o-ribosomes), also known as specialized ribosomes. In terms of reprogramming translation, o-ribosomes are especially powerful as they enable one to partially decouple translation from the native protein synthesis machinery. In particular, o-ribosomes can translate genes with altered Shine-Dalgarno (SD) sequences not recognized by host ribosomes. Because of this fact, o-ribosomes can be used to explore translational regulatory mechanisms such as coupling and also can be used to probe ribosome structure. Furthermore, o-ribosomes can be used to explore gene expression dynamics as they can be used to tune translation rates. Finally, o-ribosomes may have application in synthetic biology as they can be used to introduce new functionality within cells.

To date, o-ribosome design has either involved ad hoc or random mutagenesis-based approaches. While these approaches clearly have been successful, one question is whether a rational, computational-based strategy could be employed in design. In particular, a computational approach would enable one to explicitly explore the different elements and associated hypotheses that factor into o-ribosome design. In this work, we propose a computational strategy for designing o-ribosomes in bacteria. The basic approach in our algorithm involves enumerating all possible ASD-SD pairs and then selecting those that minimally interfere with the translation of native mRNA. To demonstrate the utility of our algorithm, we experimentally tested a number of computationally designed o-ribosomes in E. coli. In the process, we were able to test a number of hypotheses regarding o-ribosome functionality. These findings should complement existing approaches based on random mutagenesis and screening. Finally, because selection is performed on the computer, this computational approach may also be useful for designing o-ribosomes in organisms where genetic screens are more difficult to implement than in E. coli. To test this hypothesis, we also used the algorithm to design o-ribosomes in Bacillus subtilis. We were able to experimentally validate these designs, though they performed less well than their counterparts in E. coli.

251b Spatiotemporal Modulation of Biodiversity in a Synthetic Predator-Prey Ecosystem

Hao Song[1], Meagan Gray[1], and Lingchong You[2]. (1) Department of Biomedical Engineering, Duke University, 101 Science Drive, Durham, NC 27708, (2) Department of Biomedical Engineering and Institute of Genome Sciences and Policy, Duke University, 101 Science Drive, Durham, NC 27708

Biodiversity is a central element in determining the structure and functionality of natural ecological systems. Its maintenance and modulation is poorly understood due to difficulties in studying this phenomenon with natural ecosystems. To this end, we have used a synthetic predator-prey ecosystem to examine effects of cellular motility and spatial configuration on biodiversity. This synthetic ecosystem consists of two Escherichia coli populations, which communicate bi-directionally through quorum sensing signaling and regulate each other's gene expression and survival via engineered gene circuits. We found that cellular motility had a negligible impact on the biodiversity when the predator and prey cells were well-mixed in either liquid phase or soft agar. Increasing inoculation distance between the two populations on soft agar, however, caused increased biodiversity. The seemingly different roles of spatial segregation and cellular motility on biodiversity can be attributed to the existence of a critical segregation distance between the predator and prey cells, as revealed by a mathematical model that accounts for the spatiotemporal dynamics of the system. Below the critical distance, cellular motility has a negligible effect; beyond it, however, biodiversity scales almost linearly with cellular motility. Our results reveal the critical importance of the interplay between cellular motility, spatial segregation, and communication for the maintenance of biodiversity in chemical-mediated ecosystems.

251c Multiple Transcription Factor Binding Sites Can Lead to Noise-Induced Bistability In Gene Expression

Tsz-Leung To and **Narendra Maheshri**. Chemical Engineering, MIT, 25 Ames St. Room 66-558, Cambridge, MA 02139

A positive feedback loop is a common network motif in genetic regulatory networks often associated with developmental cell-fate decisions, memory, and bistable "all-or-none" responses. An often-stated "necessary" condition for bistability predicted by deterministic dynamic models is the presence of non-linear, switch-like feedback. A common molecular mechanism given to account for the non-linear response is cooperative binding of transcription factors to multiple binding sites within the promoter of a gene. Surprisingly, we find experimentally that linear feedback is capable of generating bistable gene expression with highly separated "ON" and "OFF" states. Furthermore, the bistable expression occurs only when multiple binding sites are present. In agreement with theoretical models, we show that noise in gene expression is responsible for the bistability. Using both single cell protein and mRNA measurements, we show that multiple binding sites lead to bistable gene expression by severely increasing this noise. We also propose a mechanism and model for how multiple binding sites increase gene expression noise. To our knowledge, this is the first experimental demonstration of noise-induced bistability in gene expression. Furthermore, it opens up questions as to what extent robust "all-or-none" responses are observed in settings where there is not significant evidence of non-linear feedback.

251d Forward Engineering of Synthetic Bio-Logical and Gates

Jonathan R. Tomshine[1], Kavita Iyer[1], Jennifer A. Maynard[2], and Yiannis N. Kaznessis[3]. (1) Chemical Engineering and Materials Science, University of Minnesota -- Minneapolis, 421 Washington Ave SE, Box 78, Minneapolis, MN 55455, (2) Chemical Engineering, University of Texas, Austin, 1 University Station, MC 0400, Austin, TX 78712, (3) Chemical Engineering and Materials Science, University of Minnesota, 151 Amundson Hall, 421 Washington Ave SE, Minneapolis, MN 55454

In recent years, the field of synthetic biology has produced genetic circuits capable of emulating functional paradigms commonly seen in digital electronic circuits such as bistable switches, oscillators, and logic gates. As biological systems, these functions have often been implemented as gene networks, with multiple genes under the control of multiple promoters. This work employs a detailed mechanistic-kinetic model and stochastic simulation techniques (following the work of Salis and Kaznessis: Salis, H., Sotiropoulos, V. & Kaznessis, Y. N. (2006) Multiscale Hy3S: Hybrid stochastic simulation for supercomputers. BMC Bioinformatics 7:93) as well as the techniques of in vivo molecular biology to study the potential of a synthetic, single promoter AND gate.

This proposed device consists of operator sites taken from the tetracycline operon (tetO) and the lactose operon (lacO) placed in the two flanking positions around the -35 and -10 sequences of the prokaryotic promoter and in the space between these critical sequences. Given two operator sequences and three possible sites within the promoter, all 6 potential configurations are constructed and evaluated. Since the cell line used (DH5αPro) has been configured to constitutively express both the LacI and TetR repressor proteins, these putative AND gate promoters are responsive to the commonly-used inducers IPTG and aTc. Only when the IPTG clears the LacI from its operator sites, and the aTc clears the TetR from its operator sites can expression of the attached gene commence. In this work, the promoter is coupled to a gene that produces GFP as an output signal. Its behavior is observed in vivo through fluorescence-activated cell sorting. The quantitative behavior of the AND gate phenotype is then studied both in silico and in vivo as a function of promoter topology.

The in silico model is constructed from kinetic data obtained from the literature and yields clearly-defined ON/OFF logical behavior at realistic inducer concentrations, matching the in vivo data. A non-reductionist modeling approach is pursued. That is, instead of developing the simplest possible model that can capture the behavior of the biological system, we include all the biomolecular interactions involved in transcription, translation, regulation and induction in order quantify the influence of each individual interaction on the overall behavior of the system. The models thus acquire a predictive character that renders them well suited for engineering purposes.

These models are then simulated by applying a hybrid stochastic-discrete/stochastic-continuous algorithm developed by Salis and Kaznessis. This approach correctly accounts for the behavior of scarce species (such as promoter and operator sites) which may be present in very small quantities – as small as one copy per cell – but whose behavior is critical to the performance of the overall system.

Because these models are stochastic, they generate probability distributions of phenotypes that are directly comparable to experimentally observed variation. Because they are detailed, the models provide the opportunity to gain molecular level insight and quantify previously undetermined biomolecular interactions.

It is observed in vivo that engineered promoters containing two tetO sites and one lacO site form AND gates of varying levels of fidelity and promoter activity, while those containing two lacO sites and one tetO site cannot be induced over the range of aTc and IPTG concentrations tested. This effect is examined and investigated in the in silico system as well, and a potential explanation for the effect is offered.

The incomplete repression by the weaker LacI repressor is also investigated and quantified. These studies generate a new parameter – a rate constant that describes the displacement of the LacI repressor by RNA polymerase – that was not obtained from the literature, but may prove helpful in future in silico design work, as it provides a critical link between the in vivo and in silico results.

The in silico results, coupled with in vivo data, not only identify important design degrees of freedom, but provide previously uncharacterized parameters that can be used to guide future synthetic designs with these commonly-used regulatory elements from the Lac and Tet operons. The design rules and insights elucidated in this work can easily be applied to the construction of future hybrid promoters. Additionally, the detailed stochastic-kinetic modeling methodology described in this work is more broadly applicable to the design of complex transcriptional systems – both single promoters and gene networks, provided the component parts are well characterized.

251e Multiplex Recombineering: Progress toward the Multigenic Insertion of Regulatory Elements and Potential Applications In Metabolic Engineering

Ryan T. Gill, Joseph R. Warner, and Lauren B. Andrews. Dept. of Chemical and Biological Engineering, University of Colorado, 1111 Engineering Drive, Campus Box 424, Boulder, CO 80309

Microbes can be used as biocatalysts for the production of commodity chemicals or fuels. However, industrial conditions can stress these organisms resulting in reduced production. Engineering stress tolerance in these strains can be difficult because many of the genes involved are unknown. We are currently identifying individual and/or contiguous genes involved in a stress tolerance phenotype by creating genome wide libraries where genes are present on a plasmid in increased copy or genes are "knocked out" by an interruption on the chromosome. Microarray analysis is used to identify mutants that propagate in the presence of stress and evaluate their relative fitness. These strategies have been successfully applied in our lab to identify E. coli genes involved in tolerance to acetate, succinate, 3-hydroxypropionate, napthol, ethanol, aspartate antimetabolites, and biofilm formation. However, these strategies do not identify associations of multiple, non-contiguous genes that are involved with stress tolerance. It is known that multiple perturbations in gene expression can result in greater stress tolerance. Methods are available to produce libraries of mutants where many genes are up regulated and down regulated within an individual but the identification of those genes remains difficult. We have designed and are constructing pools of synthetic DNA containing barcode tags, regulatory elements and gene homology regions that allow precise insertion upstream of almost every gene in E. coli. The barcode tags allow genome wide identification of genes containing synthetic regulatory elements in a complex population of mutants on a universal microarray. Synthetic regulatory elements that increase or decrease expression, respectively, include the PLtetO-1 promoter fused to a ribosome binding site or an "inert" sequence that replaces the native ribosome binding site and start codon. The pool of synthetic DNA is then transformed into E. coli and chromosomal insertion is catalyzed by the bacteriophage lambda-Red proteins, termed "Recombineering". Multiplex DNA synthesis and multiplex recombineering are currently being optimized. Multiplex recombineering is a potential new way to generate a library of mutants with genome wide multigenic changes in expression. Individuals in the libraries can be selected for growth in stress conditions and genes involved in tolerance identified on a barcode tag microarray. This method has the potential to create libraries where genes are up regulated and/or down regulated and mutants are identified relatively easily. Additionally, genetic diversity may be tuned by the simultaneous or sequential insertion of regulatory elements for multiple genes.

251f Engineering Microbial Production of Glucaric Acid

Tae Seok Moon, Sang-Hwal Yoon, and **Kristala L. Jones Prather**. Chemical Engineering, Massachusetts Institute of Technology, 77 Massachusetts Avenue, Room 66-425, Cambridge, MA 02139

Organic acids represent a class of compounds with both commercial interest and significant potential for biological production. Glucaric acid is one such molecule, deemed a "value-added" compound for integration into biorefineries. We have engineered strains of *E. coli* capable of producing glucaric acid from glucose in three steps. Two of the genes utilized have been previously cloned and reported in the literature. INO1 enables the production of myo-inositol from glucose, and MIOX oxidizes myo-inositol to glucuronic acid. We have cloned and characterized a third enzyme, uronate dehydrogenase, from multiple sources, that facilitates the conversion of glucuronic acid to glucaric acid. Flux towards both glucuronic and glucaric acids is ultimately limited by MIOX, whose activity is dependent upon its substrate concentration. We have used synthetic scaffolds to co-localize INO1 and MIOX, thereby increasing the local concentration of myo-inositol. We will discuss the effects of these synthetic methods on improving the performance of the biosynthetic pathway. In addition, we will present progress to-date on the construction of an alternative designed pathway towards glucuronic acid that proposes putative transformations based on the selection of enzymes primarily according to generalized enzyme reactions and not specific substrate-enzyme pairs. This latter method follows a framework of synthetic biology for "retro-biosynthetic design" of novel pathways and requires that new activities consisting of enzymes with altered substrate specificity be created.

251g Secreting Cellulases from Salmonella

Danielle Tullman Ercek and Christopher Voigt. Pharmaceutical Chemistry, University of California San Francisco, 1700 4th Street, Byer's Hall Rm 408C, MC2540, San Francisco, CA 94158

We have optimized an engineered version of the type III secretion system of Salmonella for the expression and secretion of enzymes that break down cellulose, known as cellulases. Previously we have shown that this system could be used to express and secrete spider silk monomers without adversely affecting cell growth or viability. Building on that effort, we now have developed and experimentally verified a model of the regulatory dynamics of the system, and applied the information that we learned from this model to increase the time of expression for cellulases or other genes of interest. A number of potential applications for our synthetic cellulase secretion system, including the high-throughput screening of cellulases and sets of cellulases for improved activity, will be discussed.

416a Metabolic Engineering Under Uncertainty: Efficient Sampling of Kinetic Data Space

Ljubisa Miskovic and Vassily Hatzimanikatis. Laboratory of Computational Systems Biotechnology, Ecole Polytechnique Federale de Lausanne (EPFL), BCH 3111 (Bâtiment de chimie UNIL), CH - 1015, Lausanne, Switzerland

Study of a metabolic network with respect to the sensitivity of the metabolite concentrations and the metabolic fluxes to genetic, biochemical, and environmental variations requires knowledge of the kinetic properties of the enzymes in the network. Experimentally observed data about the in vivo enzyme kinetics are subject to large variations of underlying physiochemical parameters. In this contribution, we present a computational framework that allows us to model the uncertainty in the kinetic data and to predict quantitatively the responses of metabolic networks in the presence of genetic and environmental perturbations. The proposed methodology relies on formalism from process control and metabolic control analysis and uses a Monte Carlo sampling method. In addition, it accounts explicitly for mechanistic properties of enzymes and physicochemical and thermodynamic constraints. The method employs a novel, efficient sampling procedure that allows us to simulate all possible meaningful states of a metabolic network. We demonstrate the properties of the proposed framework through a case study for the identification of rate limiting steps in the bacterial pentose phosphate pathway.

416b Metabolic Network Reconstruction and Analysis of Butanol Producing Strain Clostridium Beijerinckii

Ravali Raju, Caroline B. Milne, and Nathan D. Price. Chemical and Biomolecular Engineering, University of Illinois Urbana-Champaign, Roger Adams Laboratory, 600 South Mathews Avenue, Urbana, IL 61801

Evidence of global warming, a desire for energy security and a push for sustainability have made research in renewable liquid fuels increasingly important. Butanol is a higher carbon alcohol with lower water affinity and vapor pressure than ethanol, and hence holds great promise as a liquid fuel alternative. Although producible using the current ethanol production process, low concentrations generated during fermentation is a major challenge. Clostridium beijerinckii has the potential to enhance industrial production of butanol in an economically advantageous manner, as it is a natural hyper-butanol producing strain and has the potential to co-ferment both pentoses and hexoses present in the cellulosic biomass feedstock. The challenge with this bacterium is that its butanol production capacity still remains below the threshold needed to maximize profitability. A systems biology approach to the rational engineering of a lesser-characterized organism such as C. beijerinckii allows for exploration of phenotype modifications using a genome-scale computational model.

The successful reconstruction and modification of an organism's metabolic network for a specific industrial goal has been demonstrated in model organisms (e.g. E. coli and S. cerevisiae). In this work, the application of metabolic reconstruction and modeling for the enhancement of a particular characteristic trait is demonstrated on C. beijerinckii NCIMB 8052. Genome annotation databases (e.g. Kegg, NCBI) and published in silico models were used to compile an initial list of metabolic genes and reactions for the reconstruction, and these data were supplemented through extensive literature searches and model-inferred gap filling. The reconstruction in itself is beneficial as it not only highlights discrepancies and missing information in the existing C. beijerinckii knowledge base, but also allows for genome-scale modeling of metabolic function using various linear algebra and computational methods.

Constraint-based analysis on the metabolic model provides information needed to rationally align the natural objectives of the organism (selection pressure for growth) with the industrial objectives of the engineer (butanol production). Specifically, by using flux balance analysis and other network optimization tools, we are able to predict network behavior for different cellular objectives and environmental states. This allows for identification of key metabolic pathways whose modification may enhance butanol production. Furthermore, in silico gene knockouts and other pathway adjustment methods available for the model all serve as guides to experimentally engineer C. beijerinckii to the network state needed to make it an economically competitive fermentation organism. Demonstration of systems biology guided cellular engineering of a lesser-characterized organism will lead to more rapid modeling and engineering of other organisms, and has the potential to significantly impact the future of biotechnology.

416c In Silico Design and Analysis of Genome-Scale E. Coli Systems Via Interactive Web Application

Dong-Yup Lee, Chemical & Biomolecular Engineering, National University of Singapore, 4, Engineering Drive 4, Singapore, 117576, Singapore and Tae-Sung Jung, Bioprocessing Technology Institute, 20 Biopolis Way, #06-01, Centros, Singapore, 138668, Singapore.

As one of best characterized and well studied bacteria, Escherichia coli has been the workhorse and model organism for biochemical, genetic and metabolic engineering researches. Since last ten years, spurred by the availability of its genome-scale in silico model [1], the research on this microbe has been further advanced within the context of systems biotechnology [2]. A multitude of applications demonstrating the predictive power of the model can be found at http://gcrg.ucsd.edu/In_Silico_Organisms/E_coli/E_coli_others. Currently, several software tools are also available for facilitating the quantitative analysis, thus allowing non-expert users to conduct in silico analysis [3]. However, none of them adopted a platform-independent web-based approach that is a new direction in software development with several advantages in terms of accessibility, software updates and computational capabilities [4]. In this study, we present a web application for in silico designing, analyzing and engineering E. coli systems. It is devised and implemented using the three-tier architecture in conjunction with emerging AJAX (Asynchronous JavaScript and XML) and Java Web Start technologies, thereby leading to tremendous increase in its usability and dynamic web accessibility. As a main feature, the present system provides a user-friendly rich web interface. It allows users to virtually create and synthesize mutant strains derived from the genome-scale wildtype E. coli model and to customize pathways of interest via a graph editor. In addition, constraints-based flux analysis can be conducted for predicting metabolic fluxes and charactering the physiological states under various genetic and/or environmental conditions in a comparative way. The usefulness and functionality of the system are demonstrated by applying to E. coli DH5α strain that was grown in 2-L batch culture provided with complex nutrient media [5]. Based on the consumption and production rates of measurable metabolites, flux analysis of virtually created E. coli DH5α mutant under the web-based environment was implemented to explore the relationship among various media components in external complex media and the internal metabolic fluxes. As a result, in silico analysis revealed that the cell utilized biosynthetic precursors for the cellular growth directly from the complex media through anabolic pathways. This indicated that the cell in complex media could be functioning in energetically more efficient manner, reducing the need to produce amino acids and saving required energy.

Acknowledgment:

This work was partially supported by the Academic Research Fund (R-279-000-258-112) from the National University of Singapore.

References:

[1] Reed, J.L., Vo, T.D., Schilling, C.H. and Palsson, B.O. 2003. An expanded genome-scale model of Escherichia coli K-12 (iJR904 GSM/GPR). Genome Biol. 4(9): R54.1-R54.12.

[2] Lee, S.Y., Lee, D.-Y. and Kim, T.Y. 2005. Systems biotechnology for strain improvement. Trends Biotechnol. 23: 349-358.

[3] Lee, D.-Y., Yun, H., Park, S. and Lee, S.Y. 2003. MetaFluxNet: the management of metabolic reaction information and quantitative metabolic flux analysis. Bioinformatics 19: 2144-2146.

[4] Lee, D.-Y., Saha, R., Yusufi, F. N. K., Park, W. and Karimi, I. A. 2008. Web-based applications for building, managing and analyzing kinetic models of biological systems. submitted.

[5] Selvarasu, S., Ow, D. S.-W., Lee, S. Y., Lee, M. M., Oh, S. K.-W., Karimi, I. A. and Lee, D.-Y. 2008. Characterizing Escherichia coli DH5α growth and metabolism in complex media using genome-scale flux analysis. submitted.

416d Towards a Dynamic Systems Approach to the Synthesis and Analysis of Mammalian Metabolic Networks

Adam C. Baughman, Lealon L. Martin, and Susan T. Sharfstein. Chemical & Biological Engineering, Rensselaer Polytechnic Institute, 110 8th Street, Ricketts Building, Troy, NY 12180

We present a novel, multi-level, computational framework for the synthesis and analysis of mammalian metabolism from the point of nutrient uptake, through a complex network of catabolic (energetic) and anabolic (biosynthetic) reactions, and to the point at which a large protein product is synthesized and possibly secreted. This framework is predicated upon dynamic systems theory, and adopts a perspective of input and output relationships as determined through a unique network state. This network state is, in turn, determined through the simultaneous optimization of a universal linear network flux-balance subject to certain unique non-linear constraints. Further, we will demonstrate the ability of this framework to identify valid, yet non-intuitive, metabolic chemistry when tasked with the optimization of cellular processes relative to the maximum yield of a protein product. We use this framework to probe for bottlenecks in synthesis of large non-native proteins (i.e. monoclonal antibodies) from mammalian cell hosts. Finally, to integrate this computational effort with the experimental laboratory, we explore specific challenges in the engineering of cell lines to adopt this non-intuitive metabolic chemistry. It is our hope that such engineered cell lines will demonstrate the improved productivity predicted by our model.

416e Factor XI Is a Potential Therapeutic Target for Thrombosis That Maintains Hemostasis

Deyan Luan[1], M. Zai[2], and Jeffrey D. Varner[1]. (1) Department of Chemical and Biomolecular Engineering, Cornell University, 120 Olin Hall, Ithaca, NY 14853, (2) Chemical and Biomolecular Engineering, Cornell University, Ithaca, NY 14850

The systems biology approach, together with computational, experimental and observational inquiry, is highly relevant to drug discovery and the optimization of medical treatments. In this study, we test our working hypothesis that fXI related mechanisms are fragile in the process of blood coagulation and could be a potential therapeutic target for thrombosis treatment. A mathematical model of the blood coagulation cascade including fXI activities was built and validated against 12 published data sets generated from two cell-based models. The simulation results show that the addition of intrinsic pathway protease fXI to the TF initiated coagulation resulted in the formation of extra amount of thrombin and thrombin was reciprocally generated in a fXI-dependent way. To gauge the robustness and fragility of each interaction in the cascade, overall state sensitivity coefficients were calculated for each of the 215 model parameters over a family of random parameter sets. Analysis of the sensitivity results for the combined case disclose that fXI auto-activations are additional fragile mechanisms to fX/FXa and thrombin related interactions. Theoretical considerations and experimental data suggest that fXI/FXIa could be excellent candidates as clinical targets in treatment of thrombotic diseases. Inhibition of fXI/FXIa could potentially reduce the risk of occlusive thrombi formation without sacrificing hemostasis as deficiency of fXI results in bleeding manifestations only after trauma or surgery. When taken together, these preliminary results support our hypothesis that fXI related mechanisms are fragile in the process of blood coagulation and could be a potential therapeutic target for thrombosis treatment.

Quantitative modeling and analysis of the transforming growth factor β signaling pathway

Seung-Wook Chung, Fayth L. Miles, Carlton R. Cooper, Mary C. Farach-Carson, and Babatunde A. Ogunnaike
University of Delaware, Newark, DE, USA

Introduction

Transforming growth factor-β (TGF-β) proteins are members of a superfamily of secreted cytokines that control a diverse array of cellular processes including cell proliferation, differentiation, motility, adhesion, angiogenesis, apoptosis, and immune surveillance. The TGF-β signaling cascade begins when extracellular TGF-β binds to and brings together Type I and Type II TGF-β receptor serine/threonine kinases on the cell surface, whereby the Type II receptor phosphorylates and activates the Type I receptor. The activated Type I receptor, in turn, propagates the signal through phosphorylation of receptor-bound (R-)Smad transcription factors (Smad2/3 and Smad1/5/8). The activated R-Smads form hetero-oligomers with Smad4 and rapidly translocate into the nucleus, undergoing continuous nucleocytoplasmic shuttling by interacting with the nuclear pore complex. Once in the nucleus, the activated Smad complexes bind to specific promoters and ultimately regulate expression of target genes, generating approximately five hundred gene responses in a cell- and context-specific manner[1, 2].

The TGF-β signaling pathway has become an attractive but problematic target for oncology drug development because of its apparently paradoxical roles in tumorigenesis and metastasis. In normal and early phase tumorigenic epithelial cells, TGF-β functions as a potent tumor suppressor primarily by inducing cell cycle arrest and apoptosis. However, in the intermediate and late stages of carcinogenesis, tumor cells become resistant to the growth inhibitory effects of TGF-β and show elevated expression of TGF-β. The ligand is over-expressed in clinical cancer samples, with increasing levels correlating with poor clinical outcomes. The role of TGF-β therefore *appears* to become one of tumor promotion, apparently supporting growth, subverting the immune system, and also facilitating, invasion, epithelial to mesenchymal transition (EMT), and angiogenesis. This finding has created the widely held perception that TGF-β acts as a tumor *promoter* in advanced tumorigenesis and metastasis[3]. Although it is known that most cancer cell lines representing the entire spectrum of tumor progression have active TGF-β signaling pathways, detailed mechanisms of how a single stimulus, TGF-β, induces such a diverse array of responses during cancer progression are not known. This is primarily due to the complexity of the signaling cascade system in which a variety of signaling components changing dynamically over different time scales interact with one another. Since quantitative understanding and analysis of such a complex regulatory circuit are not possible via qualitative human intuition alone, mathematical descriptions that lead to predictive models are necessary.

To improve our understanding of complex TGF-β signaling quantitatively, we have developed a comprehensive, dynamic model of the canonical TGF-β pathway via Smad transcription factors, based on the most up-to-date information available in the literature. Through simulation and model analysis, our model provides insight into the signal-response relationship between the binding of TGF-β to its receptor at the cell surface and the activation of downstream effectors in the signaling cascade. In particular, we use the model to carry out "*in-silico* mutations" from which we generate several hypotheses regarding potential mechanisms for how TGF-β's tumor-suppressive roles may appear to morph into tumor-promoting roles.

Model Formulation

Our model incorporates the following essential molecular processes: (i) sequential activation of receptors induced by TGF-β; (ii) receptor internalization to endosomes and recycling; (iii) R-Smad phosphorylation by receptor complex; (iv) R-Smad-Smad4 complex formation in the cytoplasm and nucleus; (v) nucleocytoplasmic shuttling of Smads; (vi) dissociation and dephosphorylation of activated Smads; (vii) constitutive and ligand-induced degradation of proteins; and (viii) protein synthesis.

The components of the overall TGF-β signaling pathway as featured in our model are depicted in Figure 1. The resulting model is a system of 17 non-linear ordinary differential equations (ODEs) with 37 kinetic parameters arising from chemical reactions represented by mass action kinetics (Table 1).

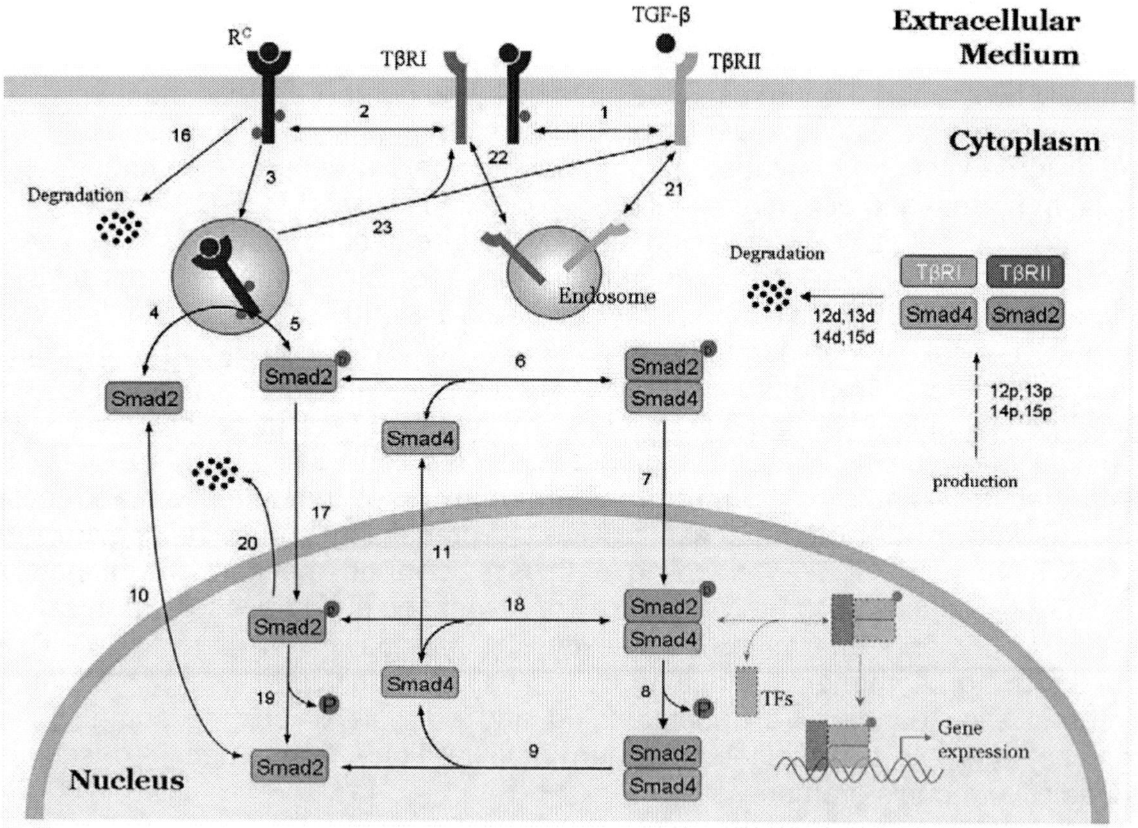

Figure 1. Schematic representation of the pathway components in the model.

Table 1. Model Equations

$$v_1 = k_{1a}[TGF\beta][RII] - k_{1d}[TGF\beta : RII]$$
$$v_2 = k_{2a}[TGF\beta : RII][RI] - k_{2d}[R^C]$$
$$v_3 = k_{3\text{int}}[R^C]$$
$$v_4 = k_{4a}[R^C_{in}][S2_{cyt}] - k_{4d}[R^C_{in} : S2_{cyt}]$$
$$v_5 = k_{5cat}[R^C_{in} : S2_{cyt}]$$
$$v_6 = k_{6a}[pS2_{cyt}][S4_{cyt}] - k_{6d}[pS2S4_{cyt}]$$
$$v_7 = k_{7imp}[pS2S4_{cyt}]$$
$$v_8 = k_{8dp}[pS2S4_{nuc}]$$
$$v_9 = k_{9d}[S2S4_{nuc}]$$
$$v_{10} = k_{10imp}[S2_{cyt}] - k_{10\exp}[S2_{nuc}]$$
$$v_{11} = k_{11imp}[S4_{cyt}] - k_{11\exp}[S4_{nuc}]$$
$$v_{12} = k_{12syn} - k_{12\deg}[RII]$$
$$v_{13} = k_{13syn} - k_{13\deg}[RI]$$
$$v_{14} = k_{14syn} - k_{14\deg}[S2_{cyt}]$$
$$v_{15} = k_{15syn} - k_{15\deg}[S4_{cyt}]$$
$$v_{16} = (k_{16\deg} + k_{16lid})[R^C]$$
$$v_{17} = k_{17imp}[pS2_{cyt}]$$
$$v_{18} = k_{18a}[pS2_{nuc}][S4_{nuc}] - k_{18d}[pS2S4_{nuc}]$$
$$v_{19}' = k_{19dp}[pS2_{nuc}]$$
$$v_{20} = k_{20lid}[pS2_{nuc}]$$
$$v_{21} = k_{21\text{int}}[RII] - k_{21rec}[RII_{in}]$$
$$v_{22} = k_{22\text{int}}[RI] - k_{22rec}[RI_{in}]$$
$$v_{23} = k_{23rec}[R^C_{in}]$$

$$\frac{d[RII]}{dt} = -v_1 + v_{12} - v_{21} + v_{23} \qquad \frac{d[TGF\beta : RII]}{dt} = v_1 - v_2 \qquad \frac{d[RI]}{dt} = -v_2 + v_{13} - v_{22} + v_{23}$$

$$\frac{d[R^C]}{dt} = v_2 - v_3 - v_{16} \qquad \frac{d[R^C_{in}]}{dt} = v_3 - v_4 + v_5 - v_{23} \qquad \frac{d[R^C_{in} : S2_{cyt}]}{dt} = v_4 - v_5$$

$$\frac{d[S2_{cyt}]}{dt} = -v_4 - v_{10} + v_{14} \qquad \frac{d[pS2_{cyt}]}{dt} = v_5 - v_6 - v_{17} \qquad \frac{d[pS2S4_{cyt}]}{dt} = v_6 - v_7$$

$$\frac{d[pS2S4_{nuc}]}{dt} = v_7 - v_8 + v_{18} \qquad \frac{d[S2S4_{nuc}]}{dt} = v_8 - v_9 \qquad \frac{d[S2_{nuc}]}{dt} = v_9 + v_{10} + v_{19}$$

$$\frac{d[S4_{nuc}]}{dt} = v_9 + v_{11} - v_{18} \qquad \frac{d[S4_{cyt}]}{dt} = -v_6 - v_{11} + v_{15} \qquad \frac{d[pS2_{nuc}]}{dt} = v_{17} - v_{18} - v_{19} - v_{20}$$

$$\frac{d[RII_{in}]}{dt} = v_{21} \qquad \frac{d[RI_{in}]}{dt} = v_{22}$$

The approach to obtain model parameter values is summarized as follows:

1. *Initial Rough Estimation*: Several kinetic parameter values were determined through an extensive literature search; some were computed using available *in vitro* experimental data; we also used physical constraints to determine others. The remaining unknown parameters were provided with initial estimates and reasonable upper and lower bounds by comparison with similar circumstances in the literature (e.g. similar steps in previous models or other signaling pathway models) and from known physical limitations (e.g. the diffusion-limited rates, 10^8-10^9 $M^{-1}s^{-1}$).

2. *Parametric Sensitivity Analysis*: To identify which parameters are the most important and which must therefore be estimated most precisely, using the set of initial estimates determined in Step 1 above, we performed local parameter sensitivity analysis to determine the effect of parametric changes on the set of five system responses of interest for which experimental measurements are available (i.e. total Smad2 in the nucleus and the cytoplasm[4], total phosphorylated Smad2 in the nucleus and the cytoplasm[4, 5], and total Smad4 in the nucleus[4]). The computations are based on the following expression for the normalized sensitivity coefficient (NSC):

$$NSC_{ij}(t) = \frac{p_j}{y_i}\frac{\partial y_i(t,\mathbf{p})}{\partial p_j}\bigg|_{\mathbf{p}}, \quad i = 1,2,\ldots,5; \quad j = 1,\ldots,37$$

where y and **p** respectively denote the system response variables and kinetic parameters. A total of 13 parameters were selected to be estimated more precisely because of their high sensitivity coefficients and/or because we had little or no confidence in their initial values.

3. *Least Squares Fitting to Data*: We fit our model predictions simultaneously to corresponding *in vitro* experimental data from the literature, via local minimization of the sum of squared residual errors:

$$\min_i R(\mathbf{p}) = \frac{1}{2}\sum (y_i(t,\mathbf{p}) - y_i^*(t))^2$$

where $y_i(t,\mathbf{p})$ and $y_i^*(t)$ denote, respectively, model predictions for a given trial of parameter values, **p**, and the corresponding experimental measurements, for each measured variable, *i*. The experimental data used for the curve-fitting are time courses of (i) total Smad2 in the nucleus[4], (ii) total Smad2 in the cytoplasm[4], (ii) total phosphorylated Smad2 in the nucleus[5], (iv) total phosphorylated Smad2 in the cytoplasm[4], and (v) total Smad4 in the nucleus[4].

4. *Identifiability*: We performed a "practical identifiability" analysis to determine whether the unknown parameters of the postulated model can be uniquely estimated from the available data, following Birtwistle *et al.*[6]. Briefly, approximate local confidence intervals for the parameter set are given by,

$$CI_i = t_{\alpha/2}^{N_t - N_p}\sqrt{\frac{S}{N_t - N_p}}\sqrt{(\mathbf{Z}^T\mathbf{Z})_{ii}^{-1}}$$

where N_t and N_p respectively denote the number of experimental data time points and the number of parameters to be estimated; $t_{\alpha/2}^{N_t - N_p}$ is the Student's *t*-distribution statistic evaluated with $N_t - N_p$ degrees of freedom, at confidence level 100(1- α)%, (with α as the "tail area probability" typically set at 0.05 to yield a 95% confidence level); *S* is the sum of squared errors, and **Z** is the model sensitivity matrix evaluated at the current parameter values. The ith parameter is said to be *practically locally identifiable* only if the magnitude of its approximate confidence interval is less than a specified tolerance i.e. $|CI_i| < \varepsilon_i$.

5. *Identifiable Parameter Estimate Refinement*: Estimated values for identifiable parameters were further refined by repeating Step 4 (local identifiability test) followed by Step 3 (local least squares estimation). After obtaining the "best" estimates of this subset of parameters, we carried out a final least squares estimation of the entire parameter set.

Results and Discussion

To help understand quantitatively which aspects of the pathway most affect the system behavior, we carried out parameter sensitivity analysis for nuclear pSmad2-Smad4 complex, which is necessary for TGF-β-induced transcriptional regulation. Figure 2 shows normalized sensitivity coefficients as a function of time for the 10 most important parameters. The most important features of this plot are summarized as follows: (1) Immediately after ligand stimulation, the output variable is strongly affected by four of this set of most sensitive parameters: in order of importance, these are k_{4a} (binding of Smad2 to active receptors), k_{3int} (internalization of receptor complexes), k_{7imp} (nuclear import of pSmad2-Smad4), and k_{2a} (complex formation of activated TβRII and TβRI). (2) On the other hand, in the mid- to longer time interval after ligand stimulation, the following parameters become more important: in order of importance, these are k_{18a} (association of nuclear pSmad2 and Smad4), and k_{18d} (dissociation of nuclear pSmad2-Smad4), k_{11exp} (nuclear export of Smad4), k_{20lid} (ligand-induced degradation of pSmad2), k_{11imp} (nuclear import of Smad4), k_{23rec} (recycling of internalized receptor complexes), k_{3int} (internalization of receptor complexes), and k_{4a} (binding of Smad2 to active receptors).

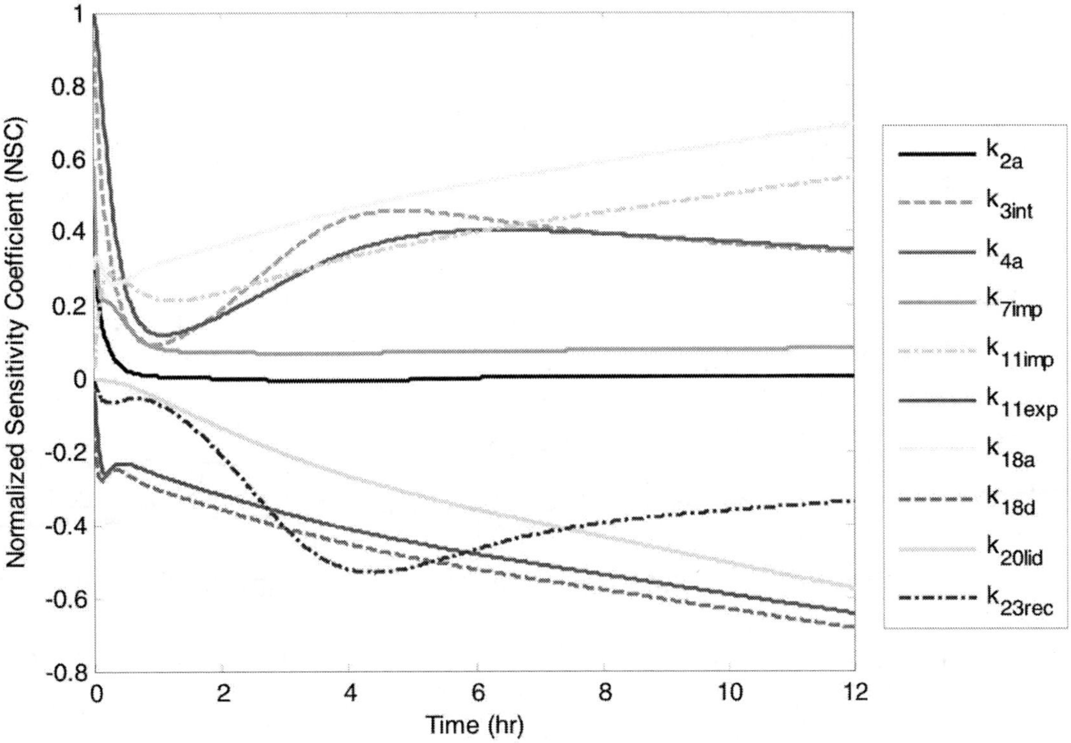

Figure 2. Model parameter sensitivities for select parameters with the greatest influence on phosphorylated Smad2-Smad4 complex in the nucleus.

These results have biologically important consequences. In particular, the increasing nature of the sensitivity coefficients of k_{18a} and k_{18d} over time shows that both the formation of pSmad2-Smad4 complex in the nucleus and its dissolution are crucial for nuclear retention of these complexes. To examine how variations in Smad complex formation in the nucleus affects nuclear accumulation of activated Smad complexes, we varied the rate of association between nuclear pSmad2 and Smad4 (k_{18a}) 10-fold. Figure 3 shows that while rapid formation of the complex between nuclear pSmad2 and Smad4 induces prolonged and enhanced nuclear accumulation of pSmad2-Smad4 complex, slow binding of pSmad2 and Smad4 leads to shortened and attenuated retention of pSmad2 complex in the nucleus. Thus, these results imply that the nuclear complex formation step plays an important role in regulating the intensity and duration of TGF-β-targeted transcriptional activities through pSmad2 complexes. More importantly, the results imply that pSmad2 complex-mediated response to TGF-β stimulation may be significantly attenuated by competitive inhibition or by interference from other nuclear molecules that also have high affinity for either pSmad2 or Smad4. This inhibitory action ultimately gives rise to a significant reduction in the rate of association between these proteins. Taken together, these results reveal that the step of complex formation between pSmad2 and Smad4 is closely associated with modulation of TGF-β-induced signal patterns.

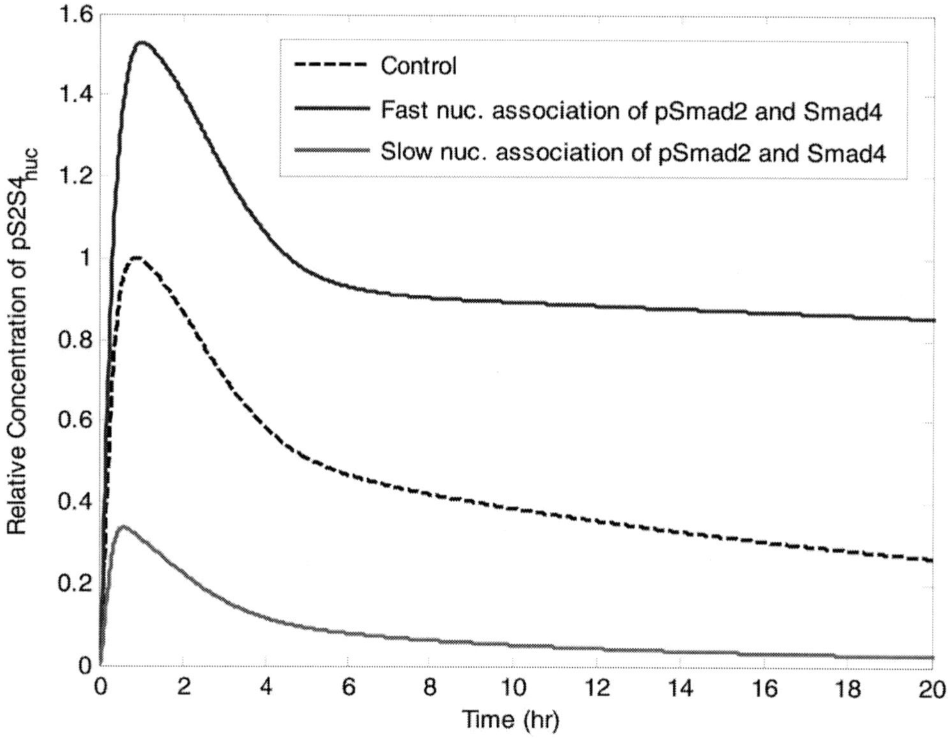

Figure 3. The effect of variations in the rate of nuclear pSmad2-Smad4 association

To help understand how cancer cells can become resistant to the tumor-suppressor effects of TGF-β, but, at the same time, remain responsive to the tumor-promoter effects, we investigated the effect on the TGF-β signaling system of abnormal alterations (e.g. mutations, deletions, downregulation, etc.) in receptors, using a 10-fold reduction in the initial levels and production rate of both Type I and Type II receptors to represent cancerous conditions. First, Figure 4 indicates that the amount of TGF-β needed to produce a saturated Smad-mediated response in cancer cells is far higher than that in healthy cells. Specifically, while the response for normal cells is essentially saturated with 1 pM of TGF-β (with higher doses producing essentially the same response), at least 10 pM of TGF-β is required before the Smad-mediated response begins to approach saturation. This is, of course, a direct effect of the reduction in the number of functional receptors in cancer cells which renders them less responsive to TGF-β stimulation. But this finding also indicates an important characteristic of cancerous cells: to elicit nuclear Smad-mediated activity generally requires *more* TGF-β than normal.

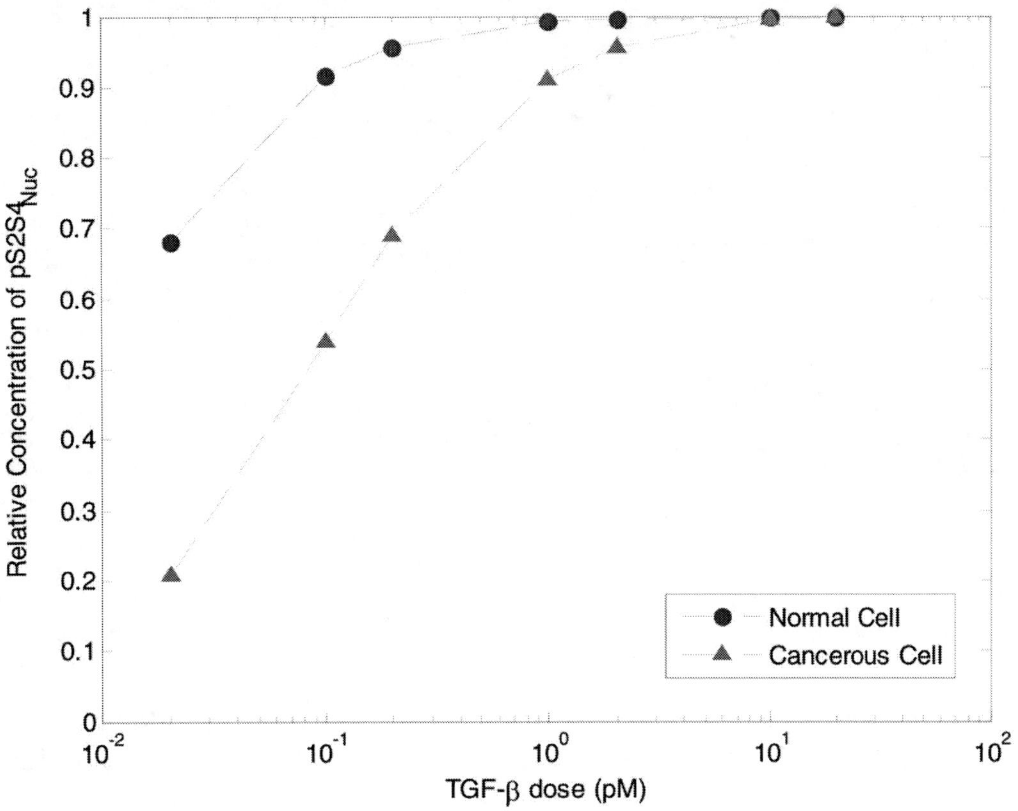

Figure 4. The effect of different concentrations of TGF-β on the maximum responses of activated nuclear Smad2-Smad4 complex in normal cells (blue circles) and in cancerous cells (red triangles).

Next, a head-to-head comparison of normal versus cancerous cell responses reveals that the sharp drop in the level of functional receptors in cancer cells leads to a marked decrease in the activity of nuclear pSmad complexes (Figure 5). Compared to the normal cell response, the peak activity of nuclear pSmad complexes in cancer cells was reduced by 65%, with the steady-state activity also remaining comparatively low. Interestingly, a reduction in the level of receptors also slowed nuclear pSmad responses. While nuclear pSmads in the normal system reached maximum activity in 55 min, their activity under a cancerous condition peaked at 86 min. Taken together, these results indicate generally that a reduction in the level of functional TGF-β receptors in cancer cells may lead to attenuated and slower TGF-β-stimulated signaling responses via Smad2. The specific implications of the model predictions in Figures 4 and 5 reveal some potentially important findings about TGF-β and cancer cells: (i) cancer cells require *higher* than normal levels of TGF-β in order to elicit significantly attenuated (and much slower) nuclear Smad-mediated activity; (ii) but even the increased levels of TGF-β will never be able to produce Smad-mediated responses that will be anywhere close to normal because of the saturation effect.

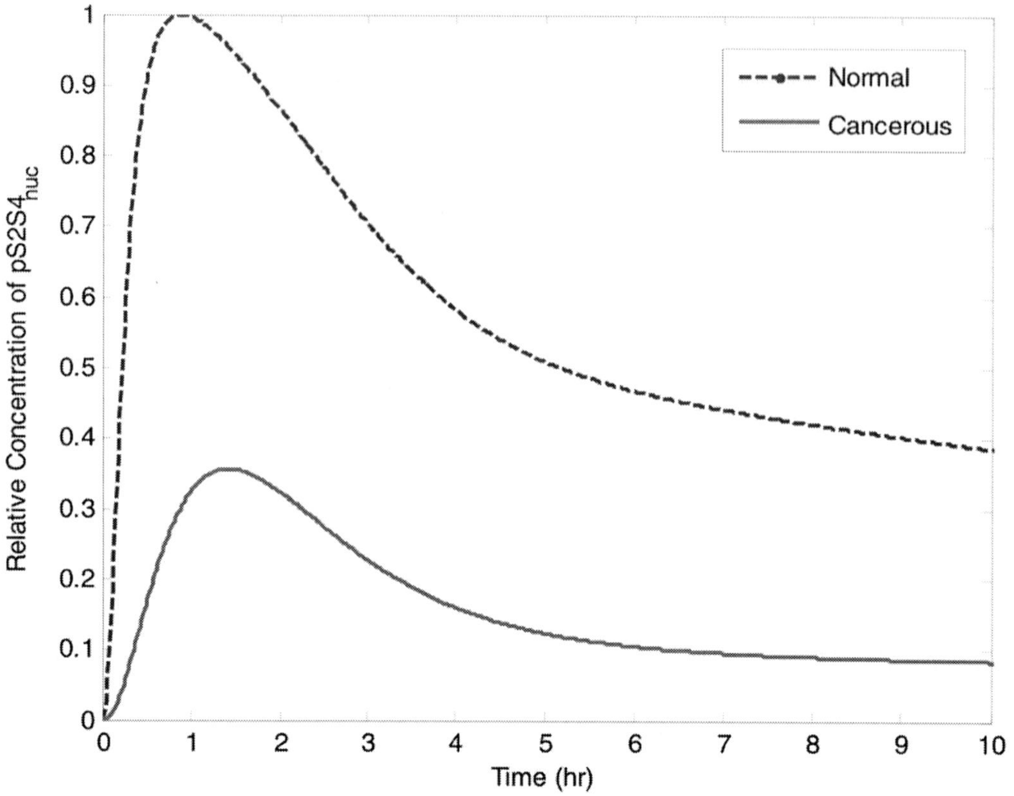

Figure 5. *In silico* mutation results: Responses of nuclear pSmad-Smad4 complex to 10-fold reduction in initial levels and protein synthesis rate constants of both Type I and Type II receptors.

As shown above via simulation, cancer cells may have attenuated TGF-β-stimulated Smad pathway responses. Cancer cells have been confirmed experimentally to be resistant to the antiproliferative effect of TGF-β, while showing typical pro-oncogenic responses. Such behavior may be explained in part by the following "threshold hypothesis": in response to TGF-β, growth-inhibitory genes require higher threshold levels of nuclear Smad activity for their expression than genes associated with pro-oncogenic and pro-metastatic effects. In other words, under normal conditions, or in the early stage of cancer progression, the antiproliferative responses to TGF-β are predominant over pro-oncogenic responses. This is because the transcriptional activity of nuclear pSmad is high enough to induce anti-growth gene expression. However, as cancer progresses, this transcriptional activity may decline significantly and thereby hardly exceed the threshold necessary for the expression of growth-inhibition genes. Meanwhile, genes related to tumor-promoting effects may be relatively insensitive to the attenuation of the transcriptional activity by Smads, so that the expression of such genes remains approximately unchanged even under cancerous conditions. As a consequence, the dominance of tumor suppressor genes over the tumor-promoter genes may be blunted in cancer cells. We believe that further investigation into differences in the temporal profiles of gene expression and thresholds of anti-growth and pro-oncogenic genes induced by TGF-β will provide some clues regarding the putative dual effects of TGF-β.

Conclusion

In this work we have presented a comprehensive computational model of the TGF-β signaling pathway that describes how an extracellular signal of TGF-β ligand is sensed by receptors and transmitted into the nucleus through intracellular Smad proteins. The model yields quantitative insight into how TGF-β-induced responses can be modulated and regulated. The model also allows us to predict possible dynamic behavior of the Smad-mediated pathway in abnormal cells, and provides clues regarding possible mechanisms for explaining the seemingly contradictory roles of TGF-β during cancer progression. We believe that incorporation of the effect of crosstalk among other important signaling cascades and detailed gene expression mechanisms into the current model will facilitate better understanding of the TGF-β paradox in cancer.

References

1. Shi, Y., and J. Massague. 2003. Mechanisms of TGF-β signaling from cell membrane to the nucleus. Cell 113:685-700.
2. Feng, X. H., and R. Derynck. 2005. Specificity and versatility in TGF-β signaling through Smads. Annu Rev Cell Dev Biol 21:659-693.
3. Pardali, K., and A. Moustakas. 2007. Actions of TGF-β as tumor suppressor and pro-metastatic factor in human cancer. Biochim Biophys Acta 1775:21-62.
4. Pierreux, C. E., F. J. Nicolas, and C. S. Hill. 2000. Transforming growth factor β-independent shuttling of Smad4 between the cytoplasm and nucleus. Mol Cell Biol 20:9041-9054.
5. Inman, G. J., F. J. Nicolas, and C. S. Hill. 2002. Nucleocytoplasmic shuttling of Smads 2, 3, and 4 permits sensing of TGFβ receptor activity. Mol Cell 10:283-294.
6. Birtwistle, M. R., M. Hatakeyama, N. Yumoto, B. A. Ogunnaike, J. B. Hoek, and B. N. Kholodenko. 2007. Ligand-dependent responses of the ErbB signaling network: experimental and modeling analyses. Mol Syst Biol 3:144.

416g Determining Regulatory Networks Governing the Differentiation of Embryonic Stem Cells to Pancreatic Lineage

Ipsita Banerjee, Massachusetts General Hospital, Harvard Medical School, 51 Blossom Street, Boston, MA 02114 and Martin Yarmush, Research, Center for Engineering in Medicine, 51 Blossom Street, Boston, MA 02114.

A lack of detailed knowledge of the gene regulatory networks governing differentiation of embryonic stem cells (ESC) significantly hinders the development of robust experimental protocols. An integrated experimental and computational approach will have significant impact in developing mature, functional cell types thus advancing the fields of stem cell therapy and regenerative medicine. We have developed a novel algorithm to capture the dynamics of gene regulatory networks involved in the differentiation of ESCs towards the pancreatic lineage from a restricted number of input data set. Our methodology can also capture the effect of external perturbations on the regulatory network. The algorithm employs a bi-level optimization method where the upper level is formulated as an integer programming problem optimizing the network topology, while the lower level deals with continuous variables optimizing the strength of the network connectivity. The input data is the time profile of gene expression during the pancreatic differentiation process. The pancreatic lineage is induced by first differentiating the ES cells to endoderms, followed by subsequent manipulations of extracellular matrix and chemical factors. The algorithm is further validated experimentally by silencing Foxa2 gene. The predicted network has an excellent agreement with the experimental data, as also with the current understanding of pancreas development. This presented integrated methodology holds great promise in developing robust in-silico differentiation protocols.

482a Smart Pooling of mRNA Samples in Microarray Experiments

Raghunandan M. Kainkaryam[1], Anna C. Gilbert[2], and Peter J. Woolf[1]. (1) Chemical Engineering, University of Michigan, Ann Arbor, MI 48109-2136, (2) Mathematics, University of Michigan, Ann Arbor, MI 48109-1043

Pooling mRNA samples in microarray experiments refers to the act of mixing mRNA from different samples and testing that mix on a microarray chip. The usual motivations for pooling include reducing biological variation, saving on chips, or lack of sufficient mRNA from samples. Recent literature suggests that pooling mRNA samples may be possible but there are a number of theoretical and experimental issues to be explored. These studies employ simple pooling strategies where similar samples (replicates) are pooled to obtain an average value of gene expression. In this talk, we propose a smart pooling strategy, called poolMC, which would test a number of different samples (n) across several chips (m, where $m < n$) and provide robust and accurate gene expression for all samples, while saving on chips used. poolMC achieves this by testing each sample multiple times in different combinations with other samples and decodes the gene expression for all genes in each sample via linear programming. poolMC is based on state-of-the-art algorithms from the field of compressed sensing. The application of poolMC to a microarray experiment requires the satisfaction of two conditions, namely linearity and sparsity. poolMC requires that the intensity measurement from a chip with pooled mRNA, for each gene, represent a linear sum of the contributions of that gene from each sample mixed on that chip. The sparsity condition provides the savings in number of chips by requiring that a gene's expression across all samples (n) being tested is mostly unchanging. Testing portions of each sample on multiple chips ensures that the decoded gene expression is robust to measurement error. We describe the theory motivating poolMC, the algorithms for experiment design and decoding of gene expression and the implementation features specific to microarrays. Finally, we demonstrate poolMC's ability to robustly and accurately quantify gene expression values for small and large sample sizes through an in-silico experiment using real gene expression data. In summary, poolMC provides a new paradigm for measuring gene expression from several samples while providing cost savings and robustness to measurement error, subject to the constraints mentioned above.

Computational Identification of Small RNAs in Clostridium acetobutylicum and Prediction of mRNA Targets

Yili Chen, Dinesh Indurthi, Shawn Jones, Eleftherios Terry Papoutsakis
University of Delaware, DE, USA

ABSTRACT

Small non-coding bacterial RNAs (sRNAs) have been found in genomes of many model organisms. Many studies show that sRNAs play important regulatory roles in a variety of cellular processes in bacteria. Clostridium acetobutylicum is a gram-positive, rodshaped anaerobe that produces acetone, butanol and ethanol through fermentation of a variety of carbon sources. It regained interest for potential use in vehicle biofuel production. However, the transcriptional regulation of *C. acetobutylicum* has not been well understood and sRNA regulation is ignored in previous studies. We predicted sRNAs and their mRNA targets in *C. acetobutylicum* ATCC 824 with various computational approaches. The non-coding sRNAs were predicted in the intergenic regions of *C. acetobutylicum* ATCC 824 genome using an integrated computational method, sRNAPredict2. The prediction was followed by Q-RT-PCR and Northern blot validation. The mRNA targets of the validated sRNAs were then predicted by searching in the genome for strong sRNA-mRNA duplexes based on sequence match and the hybrid profile prediction. In summary, 133 sRNAs were predicted, 117 on the chromosome and 16 on the plasmid. Experiments verified the expression of 7 out of 15 randomly selected putative sRNAs. The study identified a group of highly conserved sRNAs that are associated with 16S Ribosomal RNA in genomic location. The high expression level of these sRNAs suggests their potentially important regulation function in *C. acetobutylicum.*

INTRODUCTION

Small non-coding bacterial RNAs (sRNAs) play important regulatory roles in a variety of cellular processes. They are typically 50–500 nucleotides in length and nearly all sRNA species identified to date are encoded in intergenic regions (IGRs). These functional RNA molecules normally do not possess a protein-coding function. Most of them act as post-transcriptional regulators by interacting with specific mRNA targets, modulating target stability and/or translation initiation. In the past few years, new experimental strategies and computational methods have been developed demonstrating that the number of sRNAs in genomes of model organisms is much higher than was previously anticipated.

Since the discovery of first set of sRNAs in E coli by accident (1), several genome-wide methods for sRNA discovery have already been developed. Many studies combined computational searches with experimental validation of selected candidates (2,3). Using a comparative genomic screen approach, Rivas et al (3) predicted 275 sRNAs in E. coli and > 11 out of the 49 tested candidates were experimentally verified. With the availability of the increasing number of bacterial genome sequences, such strategy successfully discovered many sRNAs not only in E coli but also in many other bacteria (4-9). For example, Livny et al. (8) developed a computer program, sRNAPredict, and identified 32 putative intergenic sRNAs in V. cholera, among which 6 were verified. In addition to the identification of sRNA encoding genes, some researchers also tried to identify sRNA targets. Beside the classical genetic approaches or microarray-based target screening, much of the recent success in identifying sRNAs targets has come from the bioinformatics aided predictions (5,10-12).

Clostridium acetobutylicum (*C. acetobutylicum*) is a gram-positive, rodshaped anaerobe that produces acetone, butanol, ethanol through fermentation of a variety of carbon sources. It recently regained interest for potential use in vehicle biofuel production. Although *C. acetobutylicum* has been studied for decades (13-16), little is known about sRNAs in this microorganism. So far, there are 23 cis-regulating riboswitches reported in the Rfam database (17), but only three sRNA (tmRNA, SRP bact, 6S) were reported. This number is much smaller than E coli. For example, in E coli K12 MG1655, 42 sRNA has been reported in Rfam database.

In this study we did a computational prediction of sRNAs in *C. acetobutylicum* ATCC 824 followed by experimental validation. The sRNAs was predicted by using multiple genetic characteristics commonly associated with sRNA-encoding genes. Expression of the predicted sRNAs was examined by Q-RT-PCR and Northern blot. The study also suggested a few mRNA targets predicted with bioinformatics approach.

MATERIALS AND METHODS

1. Computational Prediction of Intergenic sRNA

sRNAPredict is a computational tool which uses coordinate-based algorithms to integrate the respective positions of individual predictive features of sRNAs and predict putative intergenic sRNAs (8). In our study the second version of this program, sRNAPredict2, was used and the predictive features were prepared as described in Livny's article (18).

The genomic sequences and genome annotations of *Clostridium acetobutylicum* ATCC 824 and its partners, *Clostridium beijerinckii, Clostridium_botulinum_A_ATCC_19397* and *Clostridium_perfringens_ATCC_13124*, were downloaded from NCBI. The tRNAs, rRNAs, previously annotated sRNAs and roboswitches were downloaded from Rfam database (17). The intergenic conserved regions between *Clostridium acetobutylicum* ATCC 824 and each of the other strains were identified with WU BLAST 2.0 (19). An E-value cut-off of $1x10^{-10}$ was applied. The putative intergenic rho-independent transcription terminators were predicted with TransTerm (20) and RNAMotif (21). The putative terminators should be no more than 20 nt downstream of the 3' end of the conserved IGRs and with confidence 96% or higher in the prediction using TransTerm. The sequence regions that likely represent conservation of RNA secondary structure were predicted with QRNA (3). A window size of 100 and a slide position of 50 were used when running the QRNA prediction.

All the predictive features were then fed into sRNAPredict2. We set the program parameters as such it only searched for sRNA sequences of 50-550 nt in the intergenic regions of *Clostridium acetobutylicum* ATCC 824 genome.

2. Computational Prediction of sRNA Targets in *C. acetobutylicum*

The reverse and complement sRNA sequence was blasted against the *C. acetobutylicum* ATCC 824 genome. The blastn parameters were set follows: Match/Mismatch = (1,-1); Gap Costs = (Existence: 2, Extension: 1); Expect threshold = 10; Word size = 7. This allows relatively low similar alignments to be found by Blast. The hybrid profiles of selected sRNA and their mRNA target candidates were then predicted with UNAFold (22,23).

3. Validation of sRNA Candidates

Bacterial strains, growth conditions, and maintenance. *C. acetobutylicum* ATCC 824 (Manassas, VA) was used as the WT strain in this study. Strains were stored at -85°C in clostridial growth medium (CGM) (24) containing 15% glycerol and revived by plating onto 2xYTG (16 g/l tryptone, 10 g/l yeast extract, 4 g/l NaCl, 5 g/l glucose, and 15 g/l agar, pH 5.8) agar-solidified plates under anaerobic conditions at 37°C. Single colonies at least 5 days old were transferred to tubes with 10 ml CGM supplemented with 80 g/l of glucose, buffered with 30 mM acetate, and adjusted to pH 7.0. The tubes were then heat shocked at 80°C for 10 min and transferred to an anaerobic incubator at 37°C.

Fermentations. An initial culture of 250 ml of CGM was inoculated from a tube grown to an A_{600} of 0.6-0.8. This initial culture was grown to an A_{600} of ~0.6 and then used to inoculate twelve subcultures of 50 ml each with a 10% inoculum. Four of these subcultures were allowed to grow unstressed, four were stressed with butyrate, and four were stressed with butanol. For the butyrate stress, 175 µl of butyric acid (Riedel-de Haën) was added at an A_{600} of 0.8, and for the butanol stress, 250 µl of 1-butanol (Fisher) was added at an A_{600} of 0.8.

RNA sampling, isolation, and cDNA generation. Samples were collected by centrifuging 3 to 10 ml of culture at 5,000xg for 10 min, 4°C and storing the cell pellet at -85°C. For the unstressed cultures, samples were taken at 6, 12, 18, and 30 hours. For the stressed cultures, samples were taken at 30 min and 1 hour after stress. RNA was isolated according to (25) with the following modification. After individual cell pellets were resuspended in 220 µl of SET buffer with lysozyme (20 mg/ml, Sigma) and proteinase K (4.55 U/ml, Roche), all cell pellets were combined and mixed together. 32 aliquots of 220 µl were then processed as described in (25). To create two pools of RNA, 25 µl of purified total RNA from each processed aliquot was combined together to give ~800 µl. For cDNA generation, the High Capacity cDNA Reverse Transcription Kit (Applied Biosystems) was used according to manufacturer's instructions. For each reaction, 2 µg of total RNA was reverse transcribed using random primers, and then diluted to 20 ng/µl.

Quantitative Reverse Transcription Polymerase Chain Reaction (Q-RT-PCR). For each Q-RT-PCR reaction, 1 µl of the reverse transcription reaction was mixed with 1 µM of each gene/sRNA specific primer (Table 1), SYBR Green PCR Master Mix (Applied Biosystems), and nuclease-free water up to 25 µl. Samples were run on an iCycler iQ5 Real-Time PCR Detection System (Bio-Rad Laboratories) following the manufacturer's instructions. Samples were run in triplicate on the same 96 well plate with three no template controls (in which instead of 1 µl of the reverse transcription reaction, 1 µl of nuclease-free water is added). Following amplification, a melt curve was performed from 55°C to 95°C at 0.5°C/10 sec, to ensure only one product was generated.

Table 1. Primers for Q-RT-PCR Verification

Gene/sRNA	Forward Primer	Reverse primer
6S 1	GAACCTACAGTTCAAACAAGGGAG	ACGATAGGTGGGTGTCCTCA
6S 3	CGCCAAGCTCTTATCTTGAACCTACAG	TCCAAAATACCGCTGCTCTT
CAC0428	AACTTCTAGTATGGCAGCTTT	AGCGTTGCAAGTACGGCTAT
CAC0681	AGTCGTATGCTCAGTTCCTGTTGA	TCCGTCTCCAATTTTTCCTG
CAC1094	GAAGATGACGACCATCCATTTGGC	GCGGCAAACATTATGAGTATTGCTTCA
CAC1322	GGAGCAGGTGTTATTGGATGCTCA	GCAGAATTTGCCTTGCTTGTTCCC
CAC2139	GATCCTAGTACAGTTAAATTCACAGC	TTGTGGTCTCCGTTTTGA
CAC2179	GCGAATATTTCGAATCCTGTGGAC	CAAAGGCTGTGGTGCTTCTTTAGG

CAC2614	TACAGAGCATGGAAGGCTATGGCT	ACCAACCCCTCGAAGTCTTT
CAC2957	CAGGTTGTAGCAGATGCAATGATGGT	CCTGCTTCAACGACTGGAAT
CAC3313	GGAGAAGAAAGAGAGGTTATCGCT	GACAAAGAGAATCCCATTTGGGAAGAC
CAP0060	GAACACATATCCAAGAGCACACACA	TGCGGGAAAGAATTTCAAAC
ch1	CAATCCGCTGTAGCAGGGTTGAAT	TTCACTTACCGCTGCTTCCT
ch6	GGACTTAATATACATGACGTAGAATC	CACCCTTTTCAAGCCATTTT
ch10	GTTCGCTGCACTAGACAGCTTAAT	CATGTTGTCACGTGAGTGCAGCAA
ch21	GAGGGTATCTAAGCTAACGACAAGAG	TCCTTCGTCTGGACTTGCTT
ch23	TCTTGATTAACTCCTCATTGACTT	TTTCCTGCATAAGTACAACAAAAA
ch25	TTCCGGGTAGCATCGCTTGAATCT	AGCCGAGTTCTGTATTCGACA
ch28	GAGCATTTCATACATAAGTTCCGTGT	TTGAATGCCCATGACCATAA
ch33	AGTTTGGAAAGGCTATTGATTT	GCTGGGCTGCCATAATAAAT
ch37	TGGGCATTATAATAGCGTCAAAGA	ACGCCACACCTAAACAATCC
ch41	TCATGCTGTAAGTGTGTGC	ACACCCTCTTTACTTATGTATT
ch43	CAGAATCGCTGTATACTGTGTAATGTA	ACCTTTCGCCAAAAGTAGGA
ch52	ATCCTTTGATAAGGAAGAGTAGCC	ATCACACCACCCTCAGCTCT
ch83	AGAGTGGCTTATAGATGTTAGT	TCGCAAATCTATTCCTCTTTCTG
pd9	AGTATCGGGAATACAAAGTCTGAT	CCTCCTGCATAAACCCCTCT
pd14	CTGCAAATAGAAATTAAGTAGGTCT	GGGTGTTACAGCACCCTATTG
SRP_bact	AATTGGGTCCCACGCAACGGAAAT	TCAGATTTATCCACGGCACA

Probe preparation for Northern analysis. Probes were designed for the detection of two predicted sRNA sequences: ch1 and ch25. Probe template was PCR amplified from one of the cDNA pools. The ch1 PCR primers were the same as the ones used for Q-RT-PCR (Table 1), and the ch25 PCR primers were 5'-GGAGTGGCCCGCTCTGCTTCCGGG-3', for the forward primer, and 5'-ACACTAAGCACGAAACCTAGTGTT-3', for the reverse primer. The probes were labeled with [α-^{32}P]dCTP (3,000 Ci/mmol) using the NEBlot Kit from New England Biolabs (Ipswich, MA), following the manufacturer's instructions, and unincorporated [α-^{32}P]dCTP was removed using illustra ProbeQuant G-50 columns (GE Healtchcare, Buckinghamshire, UK), following the manufacturer's instructions.

Northern blot and analysis. Twenty-microgram samples of total RNA were run on a 2.0% MOPS-formaldehyde agarose gels with 0.05 µg/ml ethidium bromide for 2.45 hours at 80 volts. Gels were imaged under UV light to ensure quality of RNA and to check for even loading of all lanes. Gels were rinsed twice with DEPC-treated water, incubated in 10 gel volumes of 10x SSC for 40 min, and transferred to a 0.45 µm pore-sized, positively charged membrane (Roche Applied Sciences, Indianapolis, IN) by capillary action using 10x SSC as a transfer buffer overnight. Membranes were fixed by heating at 80°C for 2 hr. Membranes were initially stained with methylene blue to confirm RNA transfer and destained by washing in 0.2x SSC and 1% SDS for 15 min at room temperature. Membranes were prehybridized with Ultrahyb Ultrasensitive Hybridization Buffer (Ambion, Austin, TX) for 2 hr at 42°C with gentle agitation. Probes were denatured by boiling for 5 min, cooled on ice for 5 min, and then added to prewarmed hybridization buffer. The prehybridization solution was poured off, the probe/hybridization buffer was added, and membranes were incubated at 42°C for 12-16 hr with gentle agitation. Finally, membranes were washed twice in 2x SSC, 0.1% SDS for 15 min at 42°C, and twice more in 0.1x SSC, 0.1% SDS for 15 min at 42°C.

RESULTS

1. Prediction of sRNAs

SRNAs are located in intergenic regions (IGRs) in the genome, we therefore searched for sRNAs in the 3,277 IGRs of *C. acetobutylicum.* Most sRNAs are conserved only among closely related species and relatively few sRNAs have been identified on the basis of their sequence homology to previously known sRNAs (26). Therefore, we analyzed the conservation of IGRs across *Clostridium acetobutylicum* ATCC 824, *Clostridium beijerinckii, Clostridium_botulinum_A_ATCC_19397* and *Clostridium_perfringens_ATCC_13124* with WU-BLAST2. To ensure the significance of the conservation, we set the BLAST E-value cut-off as 1e-10. The sRNAs were searched only in these conserved IGRs. The putative Rho-independent terminator is an important marker of the end of transcripts in bacteria. In this study, it was identified using TransTerm (20) and RNAMotif (27). A rule used in TransTerm prediction was that putative terminators should be no more than 20 nt downstream of the 3' end of the conserved IGRs and with confidence 96% or higher.

All the above features were fed into RNAPredict2, which is a computational tool that uses coordinate-based algorithms to integrate the respective positions of individual predictive features of sRNAs and predict putative intergenic sRNAs (8,28). In total, we predicted 133 sRNA in the genome of *C. acetobutylicum* ATCC 824, of which 117 were on the chromosome and 16 were on the plasmid. In the 133 predicted non-coding RNA on the chromosome, 102 were novel and 31 matched the annotated *cis*-regulatory RNA element or riboswitches on the chromosome in Rfam database (17).

2. Validation of sRNA Prediction.

In order to experimentally verify the predictions, Q-RT-PCR was performed on 15 predicted sRNA sequences, which were largely randomly selected. Their cycle threshold (Ct) values were compared to both positive and negative control sequences to determine whether they are expressed or not, and if expressed, their relative expression level (Figure 1). Three of the positive controls (6S 1, 6S 3, and signal recognition particle (SRP bact)) are noncoding RNA genes identified in almost all sequenced bacteria(29-31), while the remaining four genes (CAC0681, CAC2139, CAC1322, CAC2957) were chosen because they are always expressed during a batch culture but at different expression levels (25). Conversely, the six negative control genes were chosen because they were never above the threshold of expression during a batch culture (25). Though genes below the threshold of expression are assumed to not be expressed, these genes could still be expressed but at very low levels compared to the rest of the transcriptome. The Ct values from three replicates of both pools were averaged together and standard deviations were calculated.

Comparing the Ct values of the two controls, three regions could be identified: genes which are expressed, genes which are either lowly expressed or not expressed, and genes which are not expressed (Figure 1). Though several negative and positive control genes overlap in the middle region, the standard deviations of the negative control genes display more variance than most of the positive control genes.

Of the 15 predicted sRNA tested, 7 were definitively expressed, 5 were either lowly or not expressed, and 3 were not expressed (Figure 1). Three of the sRNA (ch10, ch52, and ch25) had Ct values comparable to or less than SRP bact, which is involved in translation and targeting of proteins to the cell membrane, indicating possible roles in global cellular function for these sRNA. Alternatively, one of these sRNA, or any of the expressed ones, could be highly upregulated during only one stage of growth or under butyrate or butanol stress, but since a pool of RNA was used, it is impossible to separate out the sRNA which are constantly expressed and only ones expressed under a certain condition. Regardless, with only three sRNA predictions being definitively not expressed, the confidence in the predictions should be quite high, and it should be noted, that the three sRNA not expressed could still be expressed under a certain stress condition not tested.

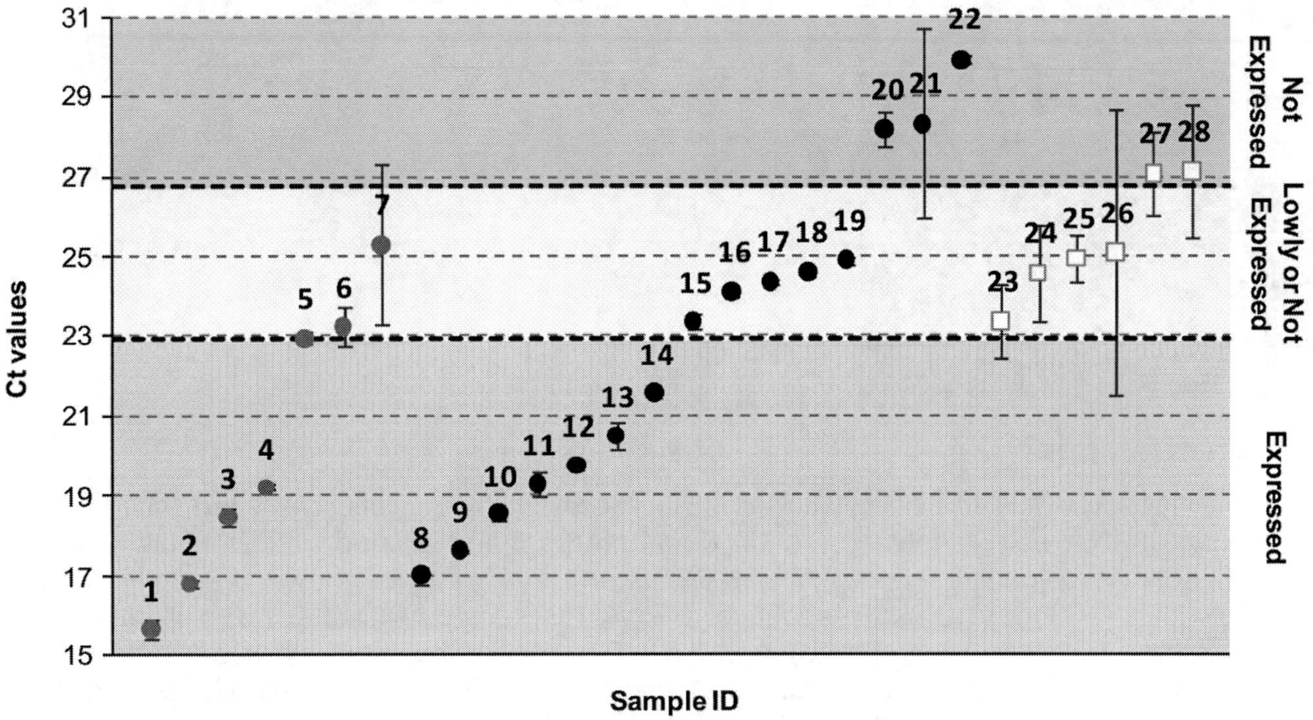

Figure 1. Q-RT-PCR cycle threshold (Ct) values of sRNA and genes tested. sRNA sequences experimentally tested (black circles, •): ch10 (8), ch52 (9), ch25 (10), ch6 (11), pd9 (12), ch1 (13), ch43 (14), ch21 (15), ch83 (16), ch33 (17), ch28 (18), ch37 (19), ch41 (20), ch23 (21), and pd14 (22). Positive controls (red cirlces, •): 6S 1 (1), 6S 3 (2), SRP_bact (3), CAC0681 (4), CAC2139 (5), CAC1322 (6), and CAC2957 (7). Negative controls (open green squares, □): CAC3313 (23), CAC0428 (24), CAC1094 (25), CAP0060 (26), CAC2179 (27), and CAC2614 (29). Genes and sRNA samples within the shaded green area are expressed, while those within the shaded yellow and red areas are either lowly expressed or not expressed. The standard deviation between 6 replicate samples is shown by the error bar for each sample.

To further validate the predictions, two sRNA were chosen for Northern blot confirmation: ch1 and ch25. Ch25 was highly expressed on the order of SRP bact, while ch1 was expressed but to a lesser extent. When the two RNA pools were probed, a single band was seen for both ch1 and ch25 (Figure 2). The band for ch1 appears to be around 400 nt (Figure 2, middle panel), which corresponds to the prediction of 314 nt for this sRNA. In contrast, the ch25 band appears to be around 300-400 nt (Figure 2, right panel), but the predicted length was only 138 nt. Thus, the predictions may not necessarily provide the correct length for the sRNA, but did predict a large enough length for identification. Also, the ch25 band appears more intense than the ch1 band, indicating that ch25 is more highly expressed than ch1, as shown in the Q-RT-PCR.

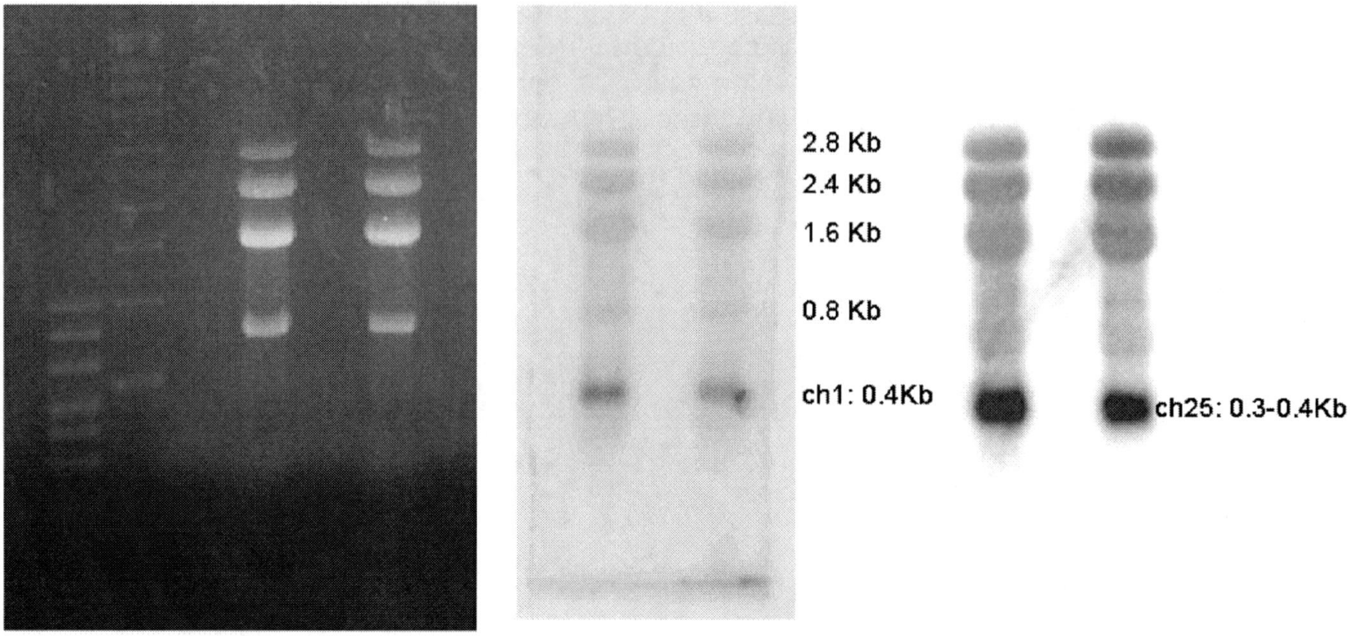

DNA ladder **ch1** **ch25**

Fig2. Detection of sRNAs of ch1 and ch25 by northern blots blots. **DNA ladder:** 18ug of pooled total RNA were resolved on a 1% denaturing gel and stained with Ethidium bromide (which acts as a loading control). Four prominent rRNA bands shown are 2.8kb (23S), 2.4kb, 1.6kb (16S) and 0.8kb respectively. RNA markers used are 0.1-1kb from Agilent (0.1, 0.2, 0.3, 0.4, 0.6, 0.8 and 1 kb), and 0.5 – 9kb RNA millennium markers from Ambion (0.5, 1, 1.5, 2, 2.5, 3, 4, 5, 6, 9 kb). **ch1:** Northern blots were probed with p32 labeled 145nt ch-1 double stranded DNA. Primers used are the same as the ones Shawn used for Q-RT PCR. Hybridization was performed at 42°C using Ambion's Ultrahyp hybridization buffer for 16-20hrs. Blots were exposed to phosphor screen for 24hrs and imaged using Typhoon imager. **ch25:** Northern blot was probed with p32 labeled 138nt ch-25 double stranded DNA. Primers used for probe synthesis.

Among the tested sRNA candidates, ch10 had the highest expression. Blast of this 360 nt sequence against the *C. acetobutylicum* ATCC 824 genome identified 10 other copies with significant similarity (E < 5e-40). All these 11 copies contain a highly conserved ~130 nt core sequence (Figure 3). It is interesting that all these copies are distributed on the inverse strand of the intergenic regions preceding rRNA-16S ribosomal RNA (16S rRNA) genes. It is not a coincidence that each of the 11 rRNA-16S rRNA genes scattering on *C. acetobutylicum* ATCC 824 genome neighbors with a highly conserved putative sRNA sequence. Q-RT-PCR validated the expression of most of these sequence copies. Five copies (ch10_1, ch10_2, ch10_3, ch10_9 and ch10_11), for which unique primers were able to be designed, showed high expression levels (Ct value < 17). In the other copies, ch10_5, ch10_6, ch10_7 and ch10_8 are identical to each other. Q-RT-PCR also showed high expression of this identical set although it cannot distinguish which one was actually expressed. In the remaining text, this identical sequence set will be named as NRS.

```
Ch10_6    168612   AGTTTGATATAATACAGATGCTAATCTGTATTCCTTTAGCTCATCGCTAAATTTATTTAA  168553
Ch10_7    336180   AGTTTGATATAATACAGATGCTAATCTGTATTCCTTTAGCTCATCGCTAAATTTATTTAA  336121
Ch10_8    341319   AGTTTGATATAATACAGATGCTAATCTGTATTCCTTTAGCTCATCGCTAAATTTATTTAA  341260
Ch10_5    163473   AGTTTGATATAATACAGATGCTAGTCTGTATTCCTTTAGCTCATCGCTAAATTTATTTAA  163414
Ch10_3    158335   AGTTTGATATAATACAGATGCTAGTCTGTATTCCTTTAGCTCATCGCTAAATTTATTTAA  158276
Ch10_1    9709     AGTTTGATATAATACAGATGCTAATCTGTATTCCTTTAGCTCATCGCTAAATTTATTTAA  9650
Ch10_4    254688   AGTTTGATATAATACAGATGCTAATCTGTATTCCTTTAGCTCATCGCTAAATTTATTTAA  254629
Ch10_10   1105276             ATACAGATGCTAATCTGTATTCCTTTAGCTCATCGCTAAATTTATTTAA  1105228
Ch10_9    331042   AGTTTGATATAATACAGATGCTAGTCTGTATTCCTTTAGCTCATCGCTAAATTTATTTAA  330983
Ch10_2    3337462    TTTGATATAATACAGATGCTAATCTGTATTCCTTTAGCTCATCGCTAAATTTATTTAA  3337519
Ch10_11   310729              ATACAGATGCTAATCTGTATTCCTTTAGCTCATCGCTAAATTTATTTAA  310681

Ch10_6    168552   ATAGAATA-AATTCTATTCAAAAGAATTGCTGGTTCTTTAATTCTCTGTTTAATTTTCAA  168494
Ch10_7    336120   ATAGAATA-AATTCTATTCAAAAGAATTGCTGGTTCTTTAATTCTCTGTTTAATTTTCAA  336062
Ch10_8    341259   ATAGAATA-AATTCTATTCAAAAGAATTGCTGGTTCTTTAATTCTCTGTTTAATTTTCAA  341201
Ch10_5    163413   ATAGAATA-AATTCTATTCAAAAGAATTGCTGGTTCTTTAATTCTCTGTTTAATTTTCAA  163355
Ch10_3    158275   ATAGAATA-AATTCTATTCAAAAGAATTGCTGGTTCTTTAATTCTCTGTTTAATTTTCAA  158217
Ch10_1    9649     ATAGAATA-AATTCTATTCAAAAGAATTGCTGGTTCTTTAATTCTCTGTTTAATTTTCAA  9591
Ch10_4    254628   ATAGAATA-AATTCTATTCAAAAGAATTGCTGGTTCTTTAATTCTCTGTTTAATTTTCAA  254570
Ch10_10   1105227  ATAGAATA-AATTCTATTCAAAAGAATTGCTGGTTCTTTAATTCTCTGTTTAATTTTCAA  1105169
Ch10_9    330982   ATAGAATA-AATTCTATTTAAAAGAATTGCTGGTTCTTTAATTCTCTGTTTAATTTTCAA  330924
Ch10_2    3337520  ATAGAATA-AATTCTATTCAAAAGAATTGCTGGTTCTTTAATTCTCTGTTTAATTTTAAA  3337578
Ch10_11   310680   ATAGAATATAATTCTATTCAAAAGAATTGCTGGTTCTTTAATTCTCTGTTTAATTTTCAA  310621

Ch10_6    168493   AGTTCAATTGTTCGCTGCACT-AGACAGCTTATATATAATATCATTTTCAGCAAGAT---  168438
Ch10_7    336061   AGTTCAATTGTTCGCTGCACT-AGACAGCTTATATATAATATCATTTTCAGCAAGAT---  336006
Ch10_8    341200   AGTTCAATTGTTCGCTGCACT-AGACAGCTTATATATAATATCATTTTCAGCAAGAT---  341145
Ch10_5    163354   AGTTCAATTGTTCGCTGCACT-AGACAGCTTATATATAATATCATTTTCAGCAAGAT---  163299
Ch10_3    158216   AGTTCAATTATCCGCTGCACT-AGACAGCTTATACATAATATCATCATCATTTAACT---  158161
Ch10_1    9590     AGTTCAATTGTTCGCTGCACT-AGACAGCTTAATTATAATATCATTTTTATTTAATC---  9535
Ch10_4    254569   AGTTCAATTGTTCGCTGCACT-AGACAGCTTATATAGAATAACATTTTTTTTCTTTG---  254514
Ch10_10   1105168  AGTTCAATTGTTTGCTGCACT-AGACAGCTTATATATATTATCATTTTGAAGAAACT---  1105113
Ch10_9    330923   AGTTCAATTGTTTGCTGCACT-AGACAGCTTAATTATATTATCA----CAACAAATTTAT  330869
Ch10_2    3337579  AGTTCAATTGTTCGCTGCACTCAGACAGCTTAATTATAATAACATCTACATTTAATC---  3337635
Ch10_11   310620   AGTTCAATTGTTCGCTGCACT-AGACAGCTTATACATATTATCATTTTCAACAAGCT---  310565

Ch10_6    168437   -TTGTCAACATCTTT-TAAAAA---CTTTTTTATTTTTAAAAGTAACAAACTTGAGACAA  168383
Ch10_7    336005   -TTGTCAACATCTTT-TAAAAA---CTTTTTTATTTTTAAAAGTAACAAACTTGAGACAA  335951
Ch10_8    341144   -TTGTCAACATCTTT-TAAAAA---CTTTTTTATTTTTAAAAGTAACAAACTTGAGACAA  341090
Ch10_5    163298   -TTGTCAACATCTTT-TAAAAA---CTTTTTTATTTTTAAAAGTAACAAACTTGAGACAA  163244
Ch10_3    158160   -TTGTCAACATCTTT-TA--------TTTAATATTTTAAAAA                    158129
Ch10_1    9534     -TTGTCAACTACTTT-T                                             9520
Ch10_4    254513   -TTGTCAACATCTTT-T                                             254499
Ch10_10   1105112  -TTGTCAACACATTTCTAAAAA---CTTTTTTATTTTTAAAAATTACAAACTTCAATCAA  1105057
Ch10_9    330868   ATTGTCAACATATTT-T                                             330853
Ch10_2    3337636  -TTGTCAACTACTTT-T                                             3337650
Ch10_11   310564   -TTGTCAACATCTTT-TAAAAAACTTTTTTTTATTTTTAAAAGCGACAAACTTGAGACAA  310507

Ch10_6    168382   CGAATACTATCATATCACATTTAATTATAAAGTCAATACATAT-TTTATAATTAT-TTTT  168325
Ch10_7    335950   CGAATACTATCATATCACATTTAATTATAAAGTCAATACATAT-TTTATAATTAT-TTTT  335893
Ch10_8    341089   CGAATACTATCATATCACATTTAATTATAAAGTCAATACATAT-TTTATAATTAT-TTTT  341032
Ch10_5    163243   CGAATACTATCATATCACATTTAATTATAAAGTCAATACATAT-TTTATAATTAT-TTTT  163186
Ch10_10   1105056  CGAATAATATCATACCACATTTATTCATATTGTCAATAAAAT-TT                 1105012
Ch10_11   310506   CGAATACTATCATACCATATTTATTAAGAATGTCAATAGTTTTATATATAAATATCTTTT  310447

Ch10_6    168324   CATGAGGATAATTTCCTCTTGGAGTAGAAAAGTGAATGCTCCATTATTAAAACATTCACT  168265
Ch10_7    335892   CATGAGGATAATTTCCTCTTGGAGTAGAAAAGTGAATGCTCCATTATTAAAACATTCACT  335833
Ch10_8    341031   CATGAGGATAATTTCCTCTTGGAGTAGAAAAGTGAATGCTCCATTATTAAAACATTCACT  340972
Ch10_5    163185   CATGAGGATAATTTCCTCTTGGAGTAGAAAAGTGAATGCTCCATTATTAAAACATTCACT  163126
Ch10_11   310446   CAT                                                           310444

Ch10_6    168264   TCTCT   168260
Ch10_7    335832   TCTCT   335828
Ch10_8    340971   TCTCT   340967
Ch10_5    163125   TCTCT   163121
```

Figure 3. Multiple sequences alignment of ch10 similar copies. The genome coordinates of the sequences in the row were annotated before and after each sequence.

3. Prediction of mRNA targets

Blast approach was applied to search target candidates of ch10 (ch10_1). 212 distinct genomic sequences were matched. Since we were looking for mRNA targets, we were particularly interested in the coding regions of the matched sequences with a reasonable length. In these 212 matched sequences, 41 are within genomic coding regions and have length greater than 34 nt. The hybrid profile of these potential mRNA targets were then predicted with UNAFold (22,23). The highest ranked target gene is listed in Table 2 based on ΔG.

The targets of NRS were predicted in the same way. Figure 4 shows its strongest sRNA–mRNA duplexes hybrid profile (ΔG = -64.4 kcal/mol) in the prediction. The genomic region of this hybrid mRNA is in the 3' end of CAC0290 (sensory transduction histidine kinase) spanning 28 nt upstream of the translation stop codon and 95 nt downstream of the annotated ORF. CAC0290 is a gene of 467 nt (328308..329015). It locates downstream of CAC0289 (Response regulator (CheY domain, HTH domain) 328308..329015) and shares the operon with CAC0289. The hybrid region on NRS is from 179 nt to 301 nt. However, this is out of the highly conserved region of the ch10 copies, which is from 56 nt to 161 nt.

Table 2. Target candidates of NRS from Blast search

Gene ID	Start .. End	Gene Description	ΔG (KCal/mol)
CAC2775	2908368..2908565	phosphohydrolase	-58.5
CAC1543	1687487..1687731	lactate dehydrogenase	-55.7
CAC0979	1126791..1126986	elongation subunit of DNA-dependent DNA polymerase	-53.2
CAC2786	2915756..2915980	hypothetical protein	-51.1
CAC1185	1336104..1336305	hypothetical protein	-50.2
CAC0356	414863..414987	putative polygalacturonase (pectinase)	-45.9
CAC3548	3744232..3744341	hypothetical protein	-44.2
CAC0946	1086612..1086775	ComE-like protein	-43.5
CAC0823	954046..954181	hypothetical protein	-41.8
CAC0052	61980..62113	hypothetical protein	-39
CAC2653	2766219..2766384	aspartate carbamoyltransferase regulatory subunit	-37.1
CAC1784	1932861..1933048	DNA uptake protein	-36.9
CAC1751	1898514..1898620	chromosome segregation SMC protein, ATPase	-35.9
CAC0691	798275..798380	hypothetical protein	-34.2
CAC2097	2194262..2194370	hypothetical protein	-34
CAC2104	2197935..2198075	general secretion pathway protein F	-33.9
CAC2064	2167897..2167999	purine nucleoside phosphorylase	-33.2
CAC1359	1504816..1504912	xylanase/chitin deacetylase	-32.4
CAC3535	3731416..3731517	Type II restriction enzyme, methylase subunit	-31.7
CAC0243	275283..275429	permease	-31.2

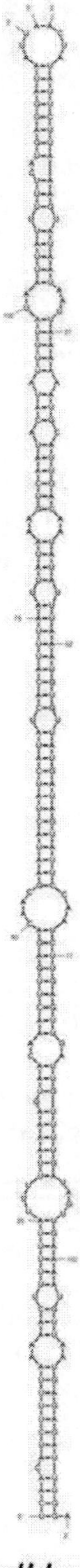

dG = -64.4 dH = -618.6 A-B

Figure 4. Predicted sRNA-mRNA duplexe for NRS. The target sequence spanning the 3'-end of CAC0290 gene (sensory transduction histidine kinase).

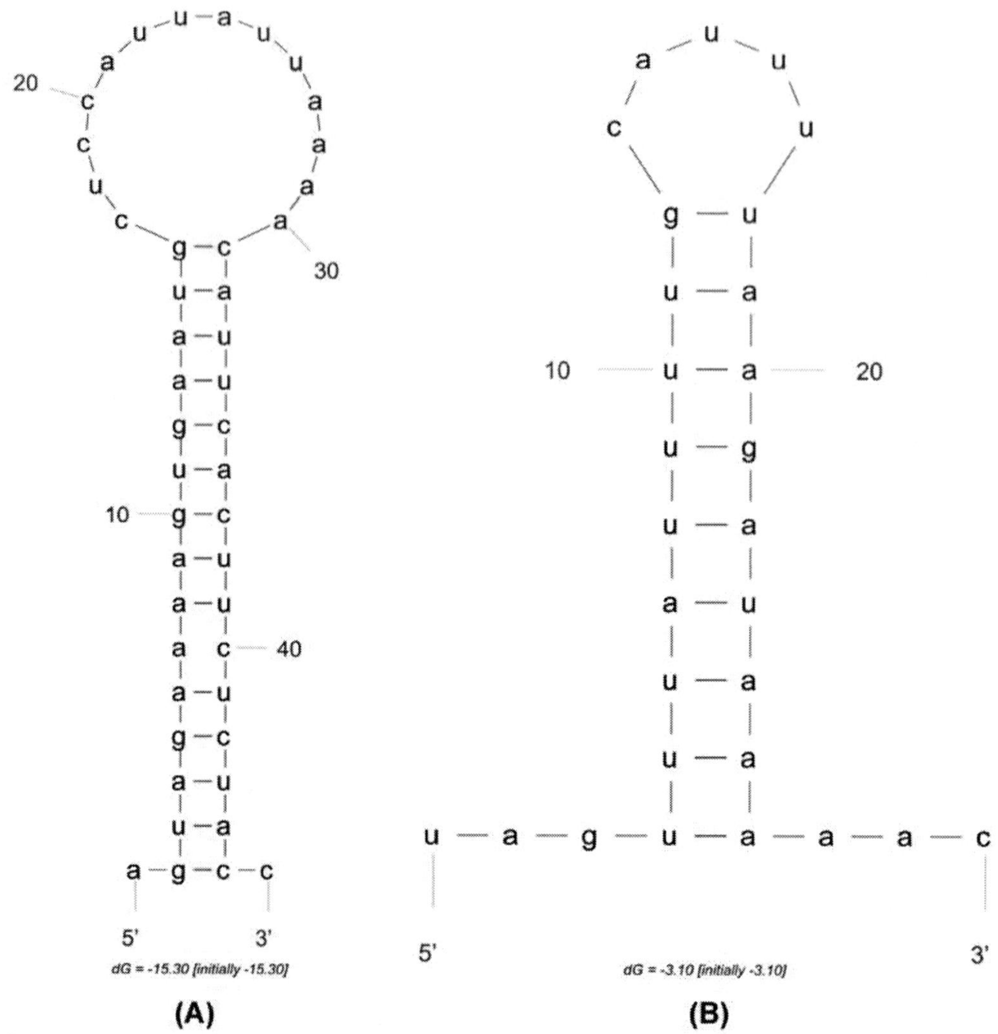

Figure 5. Potential transcript terminator of the validated ch10 sRNA copies. **A.** Loop structure near the end of NRS transcript. The loop sequence is on the inverse strand of the rho-independent terminator of 5S ribosomal RNA, which is upstream of the complementary sequence loci of NRS. **B.** Loop structure near the end of ch10_1 sRNA transcript.

DISCUSSION

This is the first study of genomic identification of sRNAs in the anaerobic bacterium Clostridium acetobutylicum. We predicted 102 novel sRNAs with size ranging from 60 nt to 500 nt. 86 of these sRNAs scatter on the chromosome and the other 16 are on the plasmid. The Q-RT-PCR and Northern blot experiment validated the expression of 7 out 15 randomly selected sRNA candidates. Although Northern blot showed that the prediction did not always capture the correct length of the real sRNA. However, it did predict a large enough length for identification.

The most interesting sRNA candidate is located on the chromosome near the 5' end of a 16S rRNA. This sRNA has a experimental validated high expression level. The sRNA sequence is highly conserved in 11 genomic loci. All these loci are located on the inverse strand of the intergenic regions preceding the 5' end of 16S rRNA genes. Q-RT-PCR verified the expression of most of these copies. The high expression levels and the consistent association of their genomic location with16S rRNA genes suggest their unknown but important functions in cells.

Despite the high similarity of these sRNA sequences, the computational prediction only caught one of the copies because no rho-independent terminator was found in the other 10 copies. However, it is noticed that NRS sequences occupies the full intergenic region, although on the inverse strand, between the genes coding for 16S rRNA and 5S rRNA. The 3' end of the 5S rRNA transcript forms a rho-independent terminator, so it is possible that the complementary sequence may also forms a weak but functional terminator structure and stop the extension of the sRNA transcription (Figure 5.A). Because this is not a classical strong rho-independent terminator, the prediction method failed to detect it and missed the sRNAs.

A strong terminator structure was also found in the ch10_1 copy, which was identified by the prediction and treated as the end of the transcript. However, Q-RT-PCR data (not included in the paper) shows that only the first half of the predicted sequence was effectively transcribed. The transcribed region is at the 5' end and contains the full conserved core sequence in the alignment (Figure 3). This means that the rho-independent terminator used to find the end of this sRNA is in fact not responsible to stop the transcription. Instead, a weak loop near the end of the transcript was found by the secondary structure prediction (Figure 5.B). But whether this weak loop is functional and determines the end of transcript needs further investigation.

The target prediction suggests a few target genes for ch10 and NRS. Given the features of how sRNA works on target mRNA, Blast is not the most suitable but still an effective approach to search for target candidates. The result provides some evidence for the potential function of the identified sRNAs.

In summary, this computational study gives new perspectives of sRNA activities in clostridium. Most importantly, it identified multiple copies of highly conserved sRNA which are located upstream of 16S rRNA. The function of these sRNAs is unknown, but their consistent genomic locations and high expression levels suggest possibilities of an import regulatory mechanism in clostridium.

REFERENCES

1. Gottesman, S. (2004) The small RNA regulators of Escherichia coli: roles and mechanisms*. *Annu Rev Microbiol*, **58**, 303-328.
2. Argaman, L., Hershberg, R., Vogel, J., Bejerano, G., Wagner, E.G., Margalit, H. and Altuvia, S. (2001) Novel small RNA-encoding genes in the intergenic regions of Escherichia coli. *Curr Biol*, **11**, 941-950.
3. Rivas, E., Klein, R.J., Jones, T.A. and Eddy, S.R. (2001) Computational identification of noncoding RNAs in E. coli by comparative genomics. *Curr Biol*, **11**, 1369-1373.
4. Wassarman, K.M., Repoila, F., Rosenow, C., Storz, G. and Gottesman, S. (2001) Identification of novel small RNAs using comparative genomics and microarrays. *Genes Dev*, **15**, 1637-1651.
5. Mandin, P., Repoila, F., Vergassola, M., Geissmann, T. and Cossart, P. (2007) Identification of new noncoding RNAs in Listeria monocytogenes and prediction of mRNA targets. *Nucleic Acids Res*, **35**, 962-974.
6. del Val, C., Rivas, E., Torres-Quesada, O., Toro, N. and Jimenez-Zurdo, J.I. (2007) Identification of differentially expressed small non-coding RNAs in the legume

endosymbiont Sinorhizobium meliloti by comparative genomics. *Mol Microbiol*, **66**, 1080-1091.
7. Panek, J., Bobek, J., Mikulik, K., Basler, M. and Vohradsky, J. (2008) Biocomputational prediction of small non-coding RNAs in Streptomyces. *BMC Genomics*, **9**, 217.
8. Livny, J., Fogel, M.A., Davis, B.M. and Waldor, M.K. (2005) sRNAPredict: an integrative computational approach to identify sRNAs in bacterial genomes. *Nucleic Acids Res*, **33**, 4096-4105.
9. Axmann, I.M., Kensche, P., Vogel, J., Kohl, S., Herzel, H. and Hess, W.R. (2005) Identification of cyanobacterial non-coding RNAs by comparative genome analysis. *Genome Biol*, **6**, R73.
10. Tjaden, B. (2008) TargetRNA: a tool for predicting targets of small RNA action in bacteria. *Nucleic Acids Res*.
11. Zhao, Y., Li, H., Hou, Y., Cha, L., Cao, Y., Wang, L., Ying, X. and Li, W. (2008) Construction of two mathematical models for prediction of bacterial sRNA targets. *Biochem Biophys Res Commun*, **372**, 346-350.
12. Tjaden, B., Goodwin, S.S., Opdyke, J.A., Guillier, M., Fu, D.X., Gottesman, S. and Storz, G. (2006) Target prediction for small, noncoding RNAs in bacteria. *Nucleic Acids Res*, **34**, 2791-2802.
13. Alsaker, K.V. and Papoutsakis, E.T. (2005) Transcriptional program of early sporulation and stationary-phase events in Clostridium acetobutylicum. *J Bacteriol*, **187**, 7103-7118.
14. Paredes, C.J., Alsaker, K.V. and Papoutsakis, E.T. (2005) A comparative genomic view of clostridial sporulation and physiology. *Nat Rev Microbiol*, **3**, 969-978.
15. Paredes, C.J., Rigoutsos, I. and Papoutsakis, E.T. (2004) Transcriptional organization of the Clostridium acetobutylicum genome. *Nucleic Acids Res*, **32**, 1973-1981.
16. Tomas, C.A., Alsaker, K.V., Bonarius, H.P., Hendriksen, W.T., Yang, H., Beamish, J.A., Paredes, C.J. and Papoutsakis, E.T. (2003) DNA array-based transcriptional analysis of asporogenous, nonsolventogenic Clostridium acetobutylicum strains SKO1 and M5. *J Bacteriol*, **185**, 4539-4547.
17. Griffiths-Jones, S., Moxon, S., Marshall, M., Khanna, A., Eddy, S.R. and Bateman, A. (2005) Rfam: annotating non-coding RNAs in complete genomes. *Nucleic Acids Res*, **33**, D121-124.
18. Livny, J. (2007) Efficient annotation of bacterial genomes for small, noncoding RNAs using the integrative computational tool sRNAPredict2. *Methods Mol Biol*, **395**, 475-488.
19. Altschul, S.F., Madden, T.L., Schaffer, A.A., Zhang, J., Zhang, Z., Miller, W. and Lipman, D.J. (1997) Gapped BLAST and PSI-BLAST: a new generation of protein database search programs. *Nucleic Acids Res*, **25**, 3389-3402.
20. Ermolaeva, M.D., Khalak, H.G., White, O., Smith, H.O. and Salzberg, S.L. (2000) Prediction of transcription terminators in bacterial genomes. *J Mol Biol*, **301**, 27-33.
21. Macke, T.J., Ecker, D.J., Gutell, R.R., Gautheret, D., Case, D.A. and Sampath, R. (2001) RNAMotif, an RNA secondary structure definition and search algorithm. *Nucleic Acids Res*, **29**, 4724-4735.
22. Markham, N.R. and Zuker, M. (2005) DINAMelt web server for nucleic acid melting prediction. *Nucleic Acids Res*, **33**, W577-581.
23. Markham, N.R. and Zuker, M. (2008) UNAFold: software for nucleic acid folding and hybridization. *Methods Mol Biol*, **453**, 3-31.
24. Wiesenborn, D.P., Rudolph, F.B. and Papoutsakis, E.T. (1988) Thiolase from Clostridium acetobutylicum ATCC 824 and Its Role in the Synthesis of Acids and Solvents. *Appl Environ Microbiol*, **54**, 2717-2722.

25. Jones, S.W., Paredes, C.J., Tracy, B., Cheng, N., Sillers, R., Senger, R.S. and Papoutsakis, E.T. (2008) The transcriptional program underlying the physiology of clostridial sporulation. *Genome Biol*, **9**, R114.
26. Livny, J. and Waldor, M.K. (2007) Identification of small RNAs in diverse bacterial species. *Curr Opin Microbiol*, **10**, 96-101.
27. Lesnik, E.A., Sampath, R., Levene, H.B., Henderson, T.J., McNeil, J.A. and Ecker, D.J. (2001) Prediction of rho-independent transcriptional terminators in Escherichia coli. *Nucleic Acids Res*, **29**, 3583-3594.
28. Livny, J., Brencic, A., Lory, S. and Waldor, M.K. (2006) Identification of 17 Pseudomonas aeruginosa sRNAs and prediction of sRNA-encoding genes in 10 diverse pathogens using the bioinformatic tool sRNAPredict2. *Nucleic Acids Res*, **34**, 3484-3493.
29. Regalia, M., Rosenblad, M.A. and Samuelsson, T. (2002) Prediction of signal recognition particle RNA genes. *Nucleic acids research*, **30**, 3368-3377.
30. Trotochaud, A.E. and Wassarman, K.M. (2005) A highly conserved 6S RNA structure is required for regulation of transcription. *Nature structural & molecular biology*, **12**, 313-319.
31. Barrick, J.E., Sudarsan, N., Weinberg, Z., Ruzzo, W.L. and Breaker, R.R. (2005) 6S RNA is a widespread regulator of eubacterial RNA polymerase that resembles an open promoter. *RNA (New York, N.Y*, **11**, 774-784.

482c Computational Analysis of Synthetic Gene/reaction Lethality In *E. coli*

Alireza Zomorrodi, Patrick F. Suthers, and Costas D. Maranas. Chemical Engineering Department, Penn State University, The Pennsylvania State University, University Park, PA 16802

Essential genes are defined as genes whose deletion is lethal. By analogy, synthetic lethals refer to pairs of non-essential genes whose simultaneous deletion is lethal (i.e., no biomass formation). One can extend the concept of synthetic lethality by considering gene groups of increasing size where only the simultaneous elimination of all genes is lethal but individual deletion are not. We developed an optimization based framework using a bilevel representation to exhaustively identify multi-gene (and by extension reaction) lethals for genome-scale models. Specifically, this framework was applied to the *i*AF1260 model of *E. coli* leading to the exhaustive identification of all double, triple and some higher-order synthetic lethals. Graphical representations of these synthetic lethals reveal a variety of motifs ranging from hub-like to highly connected sub-graphs. These results provide a birds-eye view of the redundant mechanisms available for redirecting metabolism and reveal complex patterns of gene utilization and interdependence. By analyzing the functional classifications of the genes involved (such as carbohydrate or nucleotide metabolism) we uncovered trends in connectivity within and between classifications.

482d Two-Transcript Gene Expression Classifiers In Diverse Phenotypes

Lucas Edelman, Department of Bioengineering, University of Illinois at Urbana-Champaign, 119 Roger Adams Lab, MC-712, Box C-3, 600 S. Mathews Avenue, Urbana, IL 61801 and Nathan D. Price, Department of Chemical and Biomolecular Engineering, University of Illinois at Urbana-Champaign, 119 Roger Adams Lab, MC-712, Box C-3, 600 S. Mathews Avenue, Urbana, IL 61801.

The advent of high-throughput DNA microarray technology has rendered genome-wide transcriptional analysis a standard practice in biological research. We describe here a simple classifier that reliably discriminates between binary phenotypic states with just two gene expression measurements. Rather than summing cumulative expression across many genes, we employ an algorithm which compares the relative expression of all possible pairs of genes in a set, and selects those gene pairs which exhibit an order of expression correlated strongly with phenotype- with a relative expression ratio greater or less than 1 dependent on the class of the sample. This algorithm is invariate to data normalization and generates robust, statistically significant classifiers even with low sample sizes. Microarray comparison of the gastrointestinal malignancies GIST and LMS generated a two-transcript classifier validated in the clinic and decisive for effective therapeutic planning. We demonstrate novel, highly accurate two-transcript classifiers in diverse phenotypes, including other cancers, cardiomyopathies, viral infection, and differentiation. These simple bivariate analyses represent a powerful modality for the stratification of cellular and pathological processes, and embody a robust analytical tool for clinical diagnostics.

482e Development and Analysis of Models for Cell-to-Cell Variability In Protein Expression

Marc R. Birtwistle and Babatunde A. Ogunnaike. Chemical Engineering, University of Delaware, 150 Academy St., Newark, DE 19716

Cell-to-cell variability in protein expression is a significant factor in driving divergent cell signaling responses to identical stimuli, and additionally, poses challenges for the design of engineered biological systems. Although protein expression variability has been extensively characterized experimentally in cell systems ranging from bacteria to mammals, how the basic mechanisms of protein expression give rise to the observed probability distributions still remains an open question. Through a combination of first-principles theoretical analysis, direct numerical simulations, empirical modeling, and comparison to experimental data, we find that over a wide range of conditions, the total protein distribution is well-modeled by the gamma distribution. The gamma distribution model is consistent with the experimentally reported scaling observations that protein expression noise (squared coefficient of variation) is inversely proportional to the mean protein abundance (43 proteins under 11 different environmental conditions in S. Cerevisiae, Bar-Even et al., 2006), and that the mean protein abundance is directly proportional to the standard deviation (Colman-Lerner et al., 2005). We find that the total protein distributions depend only on two parameters, θ, the mean number of proteins produced per expression burst, and the composite $k_{deg}\tau$, the product of the first-order degradation rate constant (k_{deg}) and the mean waiting time between expression bursts (τ). Our model analysis reveals that while the mean protein abundance depends on both θ and $k_{deg}\tau$, both protein expression noise and the proportionality constant between mean protein abundance and standard deviation are uniquely determined by $k_{deg}\tau$. Our results therefore suggest that protein expression noise can be manipulated independently of mean protein abundance. Additionally, we find that protein expression variance may be reduced by regulating the waiting times between expression bursts, but only when $k_{deg}\tau$ is small (< 1). The understanding gained here not only yields quantitative insight into the how the fundamental processes of protein expression and degradation contribute to protein expression variability, but can also be exploited in the design of engineered biological systems to tune protein expression noise.

References

Bar-Even, A. et al. Noise in protein expression scales with natural protein abundance. Nat Genet 38, 636-43 (2006).

Colman-Lerner, A. et al. Regulated cell-to-cell variation in a cell-fate decision system. Nature 437, 699-706 (2005).

482f A Mathematical and Computational Approach for Integrating the Major Sources of Cell Population Heterogeneity

Michail Stamatakis and Kyriacos Zygourakis. Chemical and Biomolecular Engineering Department, Rice University, Houston, TX 77005

Several approaches have been used in the past to model heterogeneity in bacterial cell populations, with each approach focusing on different source(s) of heterogeneity. However, a holistic approach that integrates all the major sources into a generic framework applicable to cell populations is still lacking.

In this work we present the mathematical formulation of a Master equation that pertains to a single cell and takes into account the major sources of heterogeneity, namely stochasticity in reaction, division, and DNA duplication. The formulation also takes into account cell growth and respects the discrete nature of the molecular contents. We further extend the framework to cell populations and develop Monte Carlo algorithms for the simulation of the stochastic processes considered here. Using this approach we demonstrate the effect of each source of heterogeneity on the overall phenotypic variability for the two-promoter system used experimentally by Elowitz et al. (2002) to quantify intrinsic versus extrinsic noise.

Elowitz, M. B., A. J. Levine, E. D. Siggia and P. S. Swain (2002). "Stochastic gene expression in a single cell." Science 297 (5584): 1183-1186

482g Optimal Tuning of Bacterial Quorum Sensing Potential

Anand Pai, Biomedical Engineering, Duke University, Pratt School of Engineering, Science Drive, Durham, NC 27708 and Lingchong You, Biomedical Engineering, Institute for Genome Sciences and Policy, Duke University, Pratt School of Engineering, Science Drive, Durham, NC 27708.

Through production and sensing of small signaling molecules, quorum sensing (QS) enables bacteria to detect changes in their density and act accordingly. A salient characteristic of QS systems is their tremendous diversity in terms of specific sensory components used in each system and their target phenotypes. Cutting across this diversity, however, the central architecture of QS is universal, which includes signal synthesis, secretion, degradation and detection. Exploiting this universality, we define a metric "sensing potential" that quantifies the ability of a single bacterium to measure the dimensions of its microenvironment. Importantly, the sensing potential provides a critical linkage between single-cell properties of a QS module and the resulting population-level outcome. In particular, tuning of the key parameters associated with QS converges in tuning of the sensing potential, which then uniquely determines whether QS is advantageous to the host bacterium. Our work provides a simple, quantitative framework to analyze the dynamics and evolution of quorum sensing.

598a Long DNA Synthesis from Oligonucleotide Mixtures by Solid Phase PCR and Assembly PCR In a Microfluidic Chip System

Woonghee Lee[1], Jean-Marie Rouillard[2], and Erdogan Gulari[2]. (1) Biomedical Engineering, The University of Michigan Ann Arbor, 2300 Hayward St., 3074 H.H. Dow Building, Ann Arbor, MI 48109, (2) Chemical Engineering, The University of Michigan Ann Arbor, 2300 Hayward St., 3074 H.H. Dow Building, Ann Arbor, MI 48109

The conventional gene synthesis methods, chemical or PCR, usually require over 2 weeks because of the separate executions of the different procedures. An integrated microfluidic chip system is designed to reduce this processing time to only 2 days with much less reaction volumes, and experimental reagent and solvent requirements. This fast high throughput gene synthesis method considerably minimizes contamination and simplifies material handling procedures. Our overall aim in this project is using the above-mentioned advantages of this system to synthesize long genes of arbitrary sequence with high purity, and cut the lead times and cost per base from the current values by at least one order of magnitude. In order to do this, four different steps are included in the microfluidic chip system: oligonucleotide synthesis and amplification on solid phase, single stranded DNA production, followed by on-chip purification, and long DNA assembly. The designed oligonucleotides to form the long DNAs are synthesized via light-directed phosphoramidite chemistry, and amplified on solid phase. The amplified products are treated to make single strands, which are later purified by on-surface hybridization using complementary probes. The purified oligonucleotides are assembled into long DNAs on chip, and amplified with polymerase chain reaction in a separate microfluidic chip chamber. Our current focus in the project is testing these individual steps in bringing the system capability to a simultaneous production level of tens of double stranded oligomers of lengths ranging from 0.2 to 1kb. In this respect, to increase the assembly yield, a single stranded DNA production method from PCR products is also under development along with a purification system which eliminates mismatch containing synthetic oligomers.

598b Identifying Biologically Significant Interactions Using Gene Set Expression Reversal Analysis

Swati Gupta[1], James A. Eddy[2], Seth Hanson[3], and Nathan D. Price[3]. (1) Biophysics and Computational Biology, University of Illinois at Urbana Champaign, MC - 195, Institute for Genomic Biology, 1206 W. Gregory Drive, Urbana, IL 61801, (2) Bioengineering, University of Illinois at Urbana Champaign, MC-195, Institute for Genomic Biology, 1206 W. Gregory Drive, Urbana, IL 61801, (3) Chemical and Biomolecular Engineering, University of Illinois at Urbana Champaign, MC - 195, Institute for Genomic Biology, 1206 W. Gregory Drive, Urbana, IL 61801

Ever evolving experimental techniques for high-throughput data generation keep data mining a major bioinformatics opportunity. Most current pathway-level comparisons of expression data are based on evaluating the enrichment of individual pathways between two sample sets (e.g. cancer vs. non-cancer or responsive vs. non-responsive to treatment). Common examples of this approach include gene ontology (GO) enrichment and Gene Set Enrichment Analysis. Here we present a microarray data analysis approach – Gene Set Expression Reversal Analysis (GSERA) – that adds to these approaches by also performing relative comparisons between all pairs of gene sets in order to uncover switches between their relative expressions between classes. The statistical significance of reversals in gene set expression levels between different biological states are determined based on their False Discovery Rates (FDR). Our program returns a sublist of top scoring gene set pairs which have the most added value as compared to either individual gene set. The user can choose between predefined metabolic/signaling pathways or dynamically define them by accessing the gene ontology database or any defined gene sets of their choosing. In cancer data sets we have studied, the program provides added information value, and very low false discovery rates, when compared with other similar programs. Our work indicates that assessing relative changes between pairs of pathways will yield significant additional insights not found using methods that consider genes or gene sets individually.

598c Genomic-Scale Transcriptional Analysis of T-Cell Activation Reveals Novel Genes and Signaling Programs in the Context of Ex Vivo Expansion

Min Wang, Department of Chemical and Biological Engineering, Northwestern University, 15 Innovation Way room288, Newark, DE 19711

Introduction. Ex vivo expanded T cells are increasingly employed in a variety of immunotherapy protocols. The success of such protocols requires large numbers of biologically active T cells. Several activation and culture protocols have been examined, but none in significant or molecular detail to allow more targeted protocol interventions. In this study, our aim was to better understand T-cell activation using a comprehensive, microarray-based comparative transcriptional analysis, in conjunction with targeted protein-level and functional analyses.

Methods. Six independent biological experiments were transcriptionally analyzed. In three biological experiments, CD3+ T cells were activated with anti-CD3/anti-CD28 Mab, cultured for 96 hours, and samples collected at 0, 4, 10, 48 and 96 hours. To cover earlier timepoints, CD4+ and CD8+ T cells of another three biological experiments were activated with anti-CD3/anti-CD28 Mab, cultured for 72 hours, and samples collected at 0, 6, 12, 48 and 72 hours. Multi-class SAM (Significance Analysis of Microarrays) with a false discovery rate <1% was used to select genes that show statistically different expression between groups. A group is defined as all the samples belonging to the same timepoint regardless of the donors. Briefly, there were 5 groups (0 hour, 4, 10, 48 and 96 hours) in the set of CD3+ experiments, and 6 groups (0 hour, 6, 12, 24, 48 and 72 hours) in the set of CD4+ experiments and CD8 + experiments. The three samples from biological experiments in each group were treated as replicates. To focus on the expression change, gene expression at each time point was compared to that of 0 hour in each experiment. Gene Ontology annotations, as curated by European Bioinformatics Institute, were retrieved from the Gene Ontology Consortium website. Interferon-γ (IFNG) and CCL20 ELISA assays, CD69, TNFSF4, CD40LG (TNFSF5), TNFRSF9, KLRD1, CD48, GZMB, CASP3, phospho-NFκB-p65, phospho-p38 (MAPK14), phospho-ERK1 (MAPK3), PUMA (BBC3), BCL2A1 protein abundance assay by flow cytometry, and AP-1 activity assay were carried out to confirm select findings of the transcriptional analysis.

Results. Comparison of the microarray-based gene expression patterns between CD3+ T cells, and the CD4+ and CD8+ subsets revealed largely conserved, but not identical, transcriptional patterns. We employed a Gene-Ontology-driven transcriptional analysis coupled with protein abundance assays in order to identify novel T-cell activation genes and cell-type-specific genes associated with the immune response. We identified potential genes involved in the communication between the two subsets (including IL23A, NR4A2, CD83, PSMB2, -8, MIF, IFI16, TNFAIP1, POU2AF1, and OTUB1) and would-be effector-function-specific genes (XCL2, SLAMF7, TNFSF4, -5, -9, CSF3, CD48 and CD244). Chemokines induced during T-cell activation, but not previously identified in T cells, include CCL20, CXCL9, -10, -11 (in all three populations), and XCL2 (preferentially in CD8+ T cells). Increased expression of other unexpected cytokines (GPI, OSM and MIF) suggests their involvement in T-cell activation with their functions yet to be examined. Differential expression of many receptors, not previously reported in the context of T-cell activation, including CCR5, CCR7, IL1R2, IL1RAP, IL6R, TNFRSF25 and TNFRSF1A, suggests their role in this immune process. Several receptors involved in TCR activation (CD3D, CD3G, TRAT1, ITGAL, ITGB1, ITGB2, CD8A and B (CD8+ T-cell specific) along with LCK, ZAP70 and TYROBP were synchronously downregulated. Members of cell-surface receptors (HLA-Ds and KLRs), none previously identified in the context of T-cell activation, were also downregulated. We also identified significantly regulated apoptotic genes in several protein families and detailed their transcriptional kinetics during the T-cell activation process. Transcription patterns of some selected genes (BCL2A1, BBC3 and CASP3) were validated at the protein level.

Conclusion. Our data and analysis provide a better understanding of T-cell activation and identify several potential targets that can be explored to benefit T-cell expansion protocols for immunotherapy applications. Although there is much known about T-cell activation, modern genomic tools provide an extraordinary opportunity to verify, extend and enrich prior knowledge, and, discover new players and processes not previously associated with T-cell activation.

598d Metagenome-Wide Metabolic Network Reconstruction of AN Acid Mine Drainage Microbial Communiy

Yu Chen, Department of Chemical Engineering, University of Michigan, 2300 Hayward Rd, 3238 H.H Dow, Ann Arbor, MI 48109 and Xiaoxia (Nina) Lin, Chemical Engineering, The University of Michigan at Ann Arbor, 2300 Hayward St., 3074 HH Dow Bldg., Ann Arbor, MI 48109.

Metagenomic sequencing has emerged as a powerful tool for the study of microbial communities, which play important roles in numerous ecosystems ranging from fungi-plant associations to human guts. In this work, we have developed a bioinformatics pipeline for generating large-scale *in silico* metabolic networks from metagenomic sequencing data and have applied it to the investigation of the metabolic capabilities of an Acid Mine Drainage (AMD) microbial community. AMD is a major source of water pollution in many mining areas and the microbial community in it is essential for its formation.

Our bioinformatics pipeline automatically reconstructs metagenome-wide metabolic networks of an entire microbial community by integrating different types of information, including metagenomic sequences and annotations, reaction/pathway databases, and organism/genome databases. We also take into count the inaccuracy and incompleteness of the used information and attempt to identify the missing components in the reconstructed network. A mixed integer linear programming (MILP) based framework using the flux balance assumption has been developed as a key step in the pipeline to reconstruct the networks and identify additional reactions which are required to fullfil assumed metabolic capabilities but are currently missing in the metagenome annotation. Next, the BLAST sequence alignment method is employed to estimate the probability of existence for all the potentially missing components in the network, which are then used to evaluate the overall uncertainty of including different sets of additional metabolic reactions in the network. By allowing alternative network reconstructions and evaluating the overall uncertainty of them, we are able to reconstruct the most probable metabolic network, capable of achieving certain pre-defined functions, such as biomass synthesis. At the same time, the enzyme candidates in the metagenome for the network gaps are identified.

We have applied the above bioinformatics pipeline to the reconstruction of the metabolic network of an AMD community which was among the first microbial communities sequenced at the metagenomic scale. This community consists of five major species. For each species, the reconstructed network contains 530 to 620 reactions. For synthesis of basic biomass components (i.e. amino acids, nucleotides and coenzymes), it requires 165-175 reactions, of which 45-80 are added without corresponding genes in the current metagenome annotation. We have identified candidate genes in the metagenome for a substantial fraction of these added reactions. We have further analyzed the reconstructed networks and discovered potential cross-feeding relationships among the species. It has been found that the biosynthesis pathways for several amino acids are very likely to be missing in *Leptospirillum sp. Group III* which is known to be the only species in the community that can fix nitrogen into ammonium required by all other community members. This hypothesis might lead to an effective treatment of AMD through blockage of the cross-feeding interactions.

598e Systems Biology Analysis of Neurodegenerative Diseases

Zheng Li[1], Ayla Ergun[1], Benjamin Wolozin[2], and James J. Collins[1]. (1) Biomedical Engineering, Boston University, 44 Cumminton St, Boston, MA 02215, (2) Depts. of Pharmacology and Neurology, Boston University School of Medicine, 715 Albany St, R-614, Boston, MA 02118

Amyotrophic lateral sclerosis (ALS) is one of the most common neurodegenerative diseases, characterized by degeneration of motor neurons in brain and spinal cord, leading to muscle atrophy and eventually death. Extensive studies have been undertaken to understand the pathogenesis of ALS. However, the underlying mechanisms of the selective degeneration of motor neurons in ALS are still incompletely understood today. It is the objective of this study to apply systems biology approaches to integrate interactome data with microarray gene expression profiles to identify the causal factors of ALS.

The interaction data including protein-protein, protein-DNA interactions of both human and mouse were collected from public databases such as NCBI. The microarray gene expression profiles of both ALS patients and mice model were obtained from GEO database. The interaction data was used to reconstruct the underlying network and gene expression profiles were applied to identify the causal pathways that were transcriptionally differentiated in ALS patients or mice model. Simulated annealing was applied to search the subnetwork with the highest score based upon transcription profile. Not surprisingly, human studies revealed an enriched pattern of cell death, neurodegeneration and inflammation. However, human samples were obtained from the patients at the end stage of the disease, which may not be informative about the causal pathways. To obtain a more complete picture of which pathways are altered within degenerating motor neurons and identify which stage of the disease these changes happen, we obtained time course gene expression profile from transgenic SOD93A mice model. Three specific stages of disease progression were evaluated, correlating with the initiation of molecular changes (60 days), the onset of muscle weakness and motor neuron loss (90 days) and late stage time point of robust motor neuron pathology (120 days). The early time point analysis indicated that the pathways such as insulin signaling, EGF signaling, PI3K-AKT, TGF signaling and antioxidant ARE-NRF2 pathway were involved in the initiation of the disease. The late stage of the mice model resembles the human studies with pathways such as apoptosis being enriched. The transcription factors and microRNAs enriched in the identified gene modules were also studied. Potential therapeutic targets were predicted and validated either with published literature or with experiments. It is also discussed how to design combinations of multiple therapeutic targets perturbations to ameliorate ALS phenotype.

598f Genomic Approach to Uncover the Nitric Oxide Response Network In Escherichia Coli

Katherine J. Chou and James C. Liao. Chemical & Biomolecular Engineering, University of California, Los Angeles, 5531 Boelter Hall, 420 Westwood Plaza, Los Angeles, CA 90095

Nitric oxide (NO) is a versatile signaling molecule in both mammalian and microbial systems. It is utilized by mammalian immune response to combat microbial invasion. To counter the NO effect, microorganisms have complex intracellular defense networks to detoxify NO. We have previously shown that under aerobic condition in minimal media, NO damages iron-sulfur protein dihydroxy acid dehydrotase (IlvD). This perturbation disrupts the biosynthesis of branch-chain amino acids and leads to E. coli bacteriostasis. However, the intracellular networks that render resistance to NO remain partially understood due to the complexity of NO chemistry, metabolism, and transport. Using a genomic approach, we systematically mapped the NO response network which confers NO-resistance in E. coli. This network involves many unknown or uncharacterized genes, which highlights the power of this approach for investigating networks beyond the well characterized genes.

598g Pathway Expression Rank Analysis (p-XRAY): A Novel Tool for Gene Set Expression Analysis

James A. Eddy, University of Illinois, Urbana, IL 61801, Donald Geman, Johns Hopkins University, Baltimore, MD 21218, and Nathan D. Price, University of Illinois Urbana-Champaign, Urbana, IL 61801.

Systems analysis of high-throughput gene expression data is useful in aiding clinical diagnoses and elucidating disease mechanisms. Due to the high dimensionality of expression data relative to the number of experimental samples (patient replicates), as well as the common need for normalization to account for microarray platform variability, it is often difficult to build robust predictive models based on single, pairs, or small clusters of genes. New approaches for expression analysis of biologically meaningful gene sets (e.g. pathways) may yield further insight into cellular disfunction and provide more informative means of differentiating complex phenotypes. We have developed a computational method, pathway Expression Rank Analysis (p-XRAY) that quantifies the regulation of gene expression within pathways in different phenotypic states. Using a priori pathway definitions, we determine a characteristic ordering of pathway genes for each phenotype based on relative levels of expression (ranks), thus normalization of gene expression values is not needed. The p-XRAY algorithm assigns scores to samples in a test set based on the conservation of this ordering in each sample, compressing expression values to a single measure for each pathway. Generating multiple sets of conservation scores for all samples, each with respect to a particular phenotype's characteristic ordering, leads to an elegant and effective classification metric – differential expression rank conservation (dERC), which performs binary classification based simply on whether the metric is positive or negative. Specifically, the dERC metric is used to find pathways where gene order is highly conserved within each of the classes being compared, but minimally conserved across all classes. We used p-XRAY with dERC to identify pathways where gene expression was highly differentially regulated between samples of gastrointestinal stromal tumors (GIST) or leiomyosarcomas (LMSs). Using these classifier pathways, we distinguished GIST and LMS with 98.6% accuracy in cross-validation. These results show p-XRAY to be a promising tool for analysis and classification of disease expression data.

667a De Novo Identification of Biomarker Proteins Using Tandem Mass Spectrometry

Peter A. DiMaggio Jr.[1], Christodoulos A. Floudas[1], Kimberly A. Salinas[2], Michael J. Hemmer[2], Jose A. Serrano[3], and Calvin C. Walker[4]. (1) Department of Chemical Engineering, Princeton University, Dept of Chemical Engineering; A215 Engineering Quadrangle, Princeton, NJ 08544, (2) US EPA Gulf Ecology Division, Gulf Breeze, FL, (3) US EPA / MED, Duluth, MN, (4) US NOAA National Marine Fisheries Service, Pascagoula, MS 39595

Many studies have shown that biological fluids contain an important number of biomarkers associated with various pathologies. For instance, there has been extensive research to identify effective biomarkers as prognostic indicators of breast cancer [1]. An effective approach for biomarker discovery is to utilize affinity-based mass spectrometry to separate proteins based on their intrinsic properties [2] using Surface Enhanced Laser Desorption/Ionization Time of Flight Mass Spectrometry (SELDI-TOF MS), where the surface of arrays attract proteins based on their affinity to chemical (hydrophobic, anionic, normal-phase, weak cation exchange, etc.) or biochemical (antibodies, enzymes, etc.) entities. Biomarkers can be discovered by comparing protein profiles for control and experimental samples to elucidate statistically significant differences in protein expression. However, the subsequent separation and identification these protein biomarkers remains a challenge.

A recent study has focused on the identification of biomarkers regarding fish species that could serve as surrogates for environmental and human health comparative studies [3]. In this work, two fish species were treated with estrogen and their plasma proteins examined for biomarkers using SEDLI-TOF-MS. The analysis revealed 13 biomarkers in a mass range of 2950 to 13000 Daltons, however, they were only able to identify one of the biomarker proteins (at 3027 Da). Some of the complications associated with the identification of these biomarkers are: (1) there exists limited sequence information for the species being studied (thus requiring de novo analysis), (2) the proteins are large in size and therefore might not fragment well using collision-induced dissocation (CID), and (3) these biomarkers must be separated from the other plasma proteins and detected in the mass spectrometer.

In this work, we present the experimental and computational techniques used to identify these protein biomarkers. Several methods for sample cleanup were applied to the plasma proteins to recover the greatest fraction of target peptides. These samples were analyzed using OrbiTrap mass spectrometry to obtain the highest accuracy data. The biomarker proteins present in the fractionated samples were analyzed in a top-down fashion, where we performed MS-N framentations using collision-induced dissociation (CID) [4] and electron-transfer dissociation (ETD) [5,6]. We present the results using both methods and highlight how these techniques can be used in a complementary fashion.

[1] Laronga C, Becker S, Watson P, Gregory B, Cazares L, Lynch H, Perry RR, Wright GL, Drake RR, Semmes OJ. SELDI-TOP serum profiling for prognostic and diagnostic classification of breast cancers. Disease Markers. 19: 229-238, 2003.

[2] Miguet L, Bogumil R, Decloquement P, Herbrecht R, Potier N, Mauvieux L, Dorsselaer AV. Discovery and identification of potential biomarkers in a prospective study of chronic lymphoid malignancies using SELDI-TOF MS. J. Proteome Res. 5: 2258-2269, 2006.

[3] Walker CC, Salinas KA, Harris PS, Wilkinson SS, Watts JD, and Hemmer MJ. A proteomic (SELDI-TOF-MS) approach to estrogen agonist screening. Toxicological Sci., 95(1):74-81, 2007.

[4] Macek B, Waanders LF, Olsen JV, and Mann M. Top-down protein sequencing and MS3 on a hybrid linear quadrupole ion trap-orbitrap mass spectrometer. Mol. Cell Proteomics, 5:949-958, 2006.

[5] Syka JEP, Coon JJ, Schroeder MJ, Shabanowitz J, Hunt DF. Peptide and protein sequence analysis by electron transfer dissociation mass spectrometry. Proc. Nat. Acad. Sci. 101(26):9528-9533, 2004.

[6] Mikesh LM, Ueberheide B, Chi A, Coon JJ, Syka JEP, Shabanowitz J, Hunt DF. The utility of ETD mass spectrometry in proteomic analysis. Biochimica. et Biophysica Acta. 1764:1811-1822, 2006.

667b Panel of Gene-Pair Relative Expression Classifiers for Stratifying Various Cancer and Non-Cancer Diseases of the Human Brain

Jaeyun Sung, Chemical and Biomolecular Engineering, University of Illinois at Urbana-Champaign, 114 Roger Adams Laboratory, MC 712, 600 South Mathews Avenue, Urbana, IL 61801 and Nathan D. Price, Chemical and Biomolecular Engineering, University of Illinois, 119 Roger Adams Lab, MC-712, 600 S. Mathews Ave., Urbana, IL 61801.

One of the main challenges in cancer therapy is the accurate diagnosis of tumor class and progression grade, which can be difficult because of morphologically indistinguishable tissue representation and inter-observer variability. In order to address this issue, cancer diagnosis is moving from classification based on microscopic observations of histologic specimens alone, to a method based on molecular criteria by using microarray gene expression profiling. One classification scheme based on gene expression microarrays is the k-Top Scoring Pairs (k-TSP) approach, a supervised machine-learning method based on the concept of finding gene-pairs that exhibit relative expression reversals, which can furthermore serve as classifiers to distinguish two different classes of disease. The goal of this study was to identify gene-pair relative expression decision rules that can distinguish seventeen different classes and progression grades of cancerous and non cancerous human brain diseases. By extending the binary classification k-TSP method into a series of multi-class decomposition schemes (1-vs-1, 1-vs-others) on an extensive collection of publicly available gene expression microarray data, we have accumulated all the candidate gene pairs from a training set into a panel of classifier markers for differentiating disease class and grade. The results show that our gene panel distinguishes each class and grade from the other sixteen classes at high accuracy using leave-one-out cross validation and also on a test set, proving to be a potential strategy for stratifying patients based on gene expression signatures. This panel of gene-pair relative expression classifiers is the first to be built for the stratification of such a large number of different brain diseases and grades, and can serve as a highly accurate diagnostic tool in combination with traditional histopathological techniques.

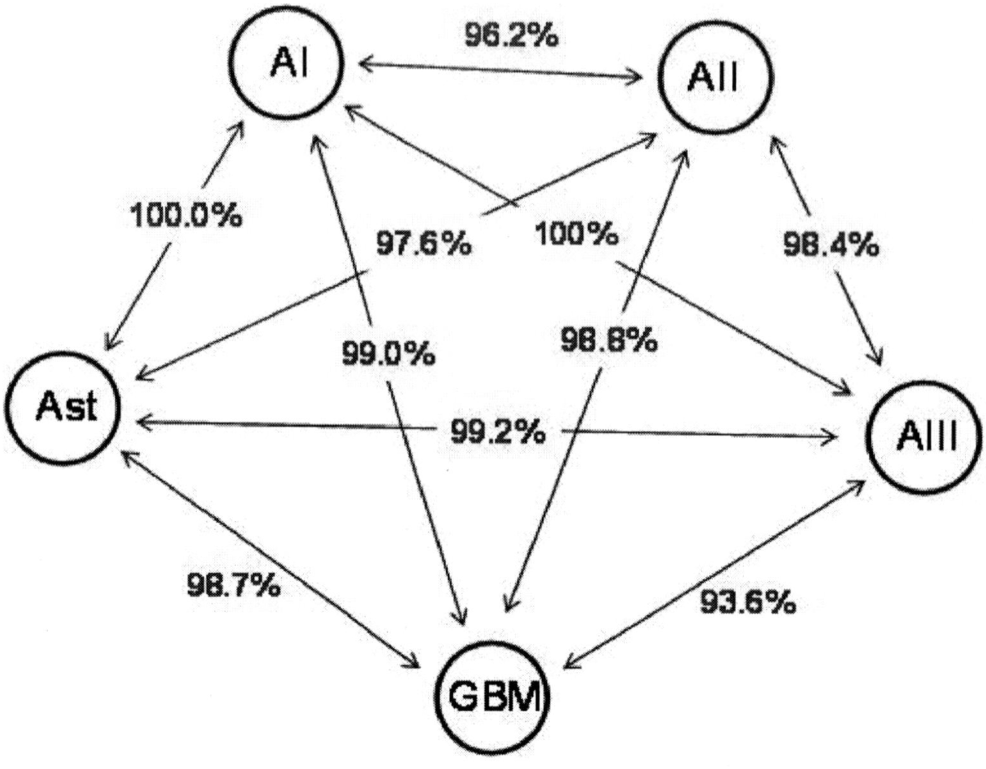

Figure 1. Accuracy of designed molecular classifiers in separating normal astrocytes (Ast) from grade I astrocytoma (AI), grade II astrocytoma (AII), grade III astrocytoma (AIII), and grade IV astrocytoma, or glioblastoma (GBM). The accuracies shown reflect pairwise comparisons for each stage of progression, and are estimates based on future cases using the leave-one-out cross-validation error estimation technique

667c Signal Transduction Networks In Cancer: Quantitative Parameters Influence Network Topology

David Klinke, Department of Chemical Engineering, West Virginia University, Morgantown, WV 26506-6102

The collective understanding of the flow of information within a cell is summarized by canonical cell signaling pathways. Analysis of expression profiling in cancer is typically done in the context of these canonical pathways. However, activation of non-canonical pathways provides a mechanism for a cell's acquisition of the hallmark characteristics of cancer. Translating basic experimental data to understand aberrant signaling networks also presents a significant challenge for the rational design of new therapies for cancer. The objective of this study was to assess the existence of non-canonical signaling pathways in breast cancer and to provide a computational framework for interpreting basic science data using concepts from reaction pathway analysis. A mathematical model of the early signaling events associated with the ErbB1 receptor was developed. The model was calibrated and validated against existing data in the literature and used as a platform to explore the activation of non-canonical signaling pathways in breast cancer cells. Protein-protein interaction data and patterns of protein expression among in vitro cell models for breast cancer were integrated within this computational framework. The cell models exhibited a broad diversity in patterns of protein expression. The phosphorylation of ErbB1 was predicted to activate Irs-1, a non-canonical interaction partner, to a similar extent as the primary canonical pathways that correspond to Shc and Grb2 interaction with ErbB1. Selecting an appropriate assay and a cellular context are shown to be critical factors for observing the protein-protein interactions responsible for non-canonical activation.

667d Unifying Flux Balance and Network-Based Pathway Analyses

Ilie Fishtik, Chemical Engineering, Worcester Polytechnic Institute, 100 Institute Road, Worcester, MA 01609

Flux balance and network-based pathway analyses are theoretical tools aimed to find optimal steady state flux distributions in a metabolic network subject to additional constraints on the rates of the reaction steps. Although mathematically these methods are rigorous there are several physicochemical aspects, mainly interpretational, that are questionable and misleading. Most of these are due to the formal disconnection between the flux balance and network-based pathway analyses. The conventional flux balance and network-based pathway analyses are reformulated in terms of reaction routes (RRs), a theoretical framework that has been developed by Horiuti over 50 years ago. Not only the theory of RRs provides the most general and rigorous definition of a pathway but it also relates the steady state rates of the reaction steps with the rates along RRs or pathways. In this presentation we show how the simple relation between the steady state rates of the reaction steps and the rates along RRs (fluxes) established by Horiuti may be used to eliminate the FBA constraints. This new approach is in fact a way to unify the flux balance and pathway analyses.

667e Mapping Condition Dependent Regulation of Lipid Metabolism

Michael C. Jewett, Genetics, Harvard Medical School, 77 Avenue Louis Pasteur, New Research Building 238, Boston, MA 02115, Christopher T. Workman, Center for Biological Sequence Analysis, Technical University of Denmark, Building 208, Kgs. Lyngby, DK-2800, Denmark, and Jens Nielsen, Chemical and Biological Engineering, Chalmers University of Technology, Kemivägen 10, Gothenburg, 412 96, Sweden.

Lipid metabolism, organization, and transport have been implicated in numerous human diseases including atherosclerosis, diabetes, and cancer, among others. Saccharamoyces cerevisiae is a model system for understanding the molecular background leading to possible human disease states. To globally map the organization and interactions of cellular networks that control lipid metabolism in S. cerevisiae, we applied a systems approach that integrated observations across multiple cellular component layers (mRNAs, lipids, metabolites, fluxes) and data from protein interaction databases with metabolic network topology. Correlation analysis revealed causal relationships and network motifs among significantly changing genes-lipids/metabolites and transcription factors with enriched target sets. Although biosynthetic metabolic genes were enriched relative to the genome, only 6 of 2351 significant correlations involve direct connections between genes and lipids/metabolites expected from a bipartite representation of a genome scale metabolic model. The correlation analysis was refined to answer the question of whether or not the biosynthetic genes connected to metabolites/lipids in the bipartite metabolic map are significantly correlated. We show that at a global level there is a high correlation between lipid levels in the cell and transcription of genes involved in lipid biosynthesis, pointing to a high degree of transcriptional regulation of metabolic fluxes towards biosynthesis of structural lipids. This is in contrast to the biosynthesis of amino acids where there is both transcriptional regulation and regulation at the metabolic level (both hierarchical and metabolic regulation). Our results contribute to the emerging view that the relationships between mRNA expression, metabolite and lipid levels, and fluxes are complex. Moreover, they demonstrate strategies for using metabolic network topology to upgrade the information content in high-throughput measurements.

667f A Functional Genomics Analysis of Central Carbon Metabolism Evolution In Yeasts

Mark P. Styczynski, Dawn A. Thompson, Jenna Pfiffner, Courtney French, and Aviv Regev. Broad Institute of MIT and Harvard, 7 Cambridge Center, Room 6175-X, Cambridge, MA 02142

Metabolic processes within a cell produce the energy and small molecule building blocks necessary for all cellular functions, and central carbon metabolism – the key metabolic module governing energy sources – is the cornerstone of these processes. Despite this key role, we understand surprisingly little about the function and evolution of central carbon metabolism on a systems scale. For example, how did the respiro-fermentative metabolic state (common to log-phase *Saccharomyces cerevisiae* and to many cancer cells) evolve and what are its adaptive advantages? In this project we perform a comparative experimental and computational study of the function, regulation, and evolution of central carbon metabolism in thirteen *Ascomycota* fungi spanning 300 million years of evolution. Using advanced metabolomics methods, we systematically measure metabolite concentrations and fluxes under a variety of metabolic and genetic perturbations in each species. We can then integrate these metabolic profiles with expression profiles collected in our lab under the same conditions and identify functional metabolic modules of co-varying metabolites and genes in each species. We use a novel computational approach to identify orthologous modules across species and reconstruct their evolution. The evolutionary reconstruction predicts conserved functional entities as well as major evolutionary changes in metabolic function and regulation. Our evolutionary model will be validated and refined by deleting predicted key genes in all relevant species in our model and using the same metabolomics-based approach to profile their responses. This innovative comparative metabolomic approach will provide insights into the function and evolution of distinct metabolic strategies and establish a new systems-level approach for the study of metabolic systems.

AUTHOR INDEX

AUTHOR INDEX

American Institute of Chemical Engineers
3 Park Avenue
New York, New York 10016-5991

Advanced Membrane Separation for Sustainability 2013

Topical Conference at the 2013 AIChE Annual Meeting: Global Challenges for Engineering a Sustainable Future

San Francisco, California, USA
3-8 November 2013

ISBN: 978-1-63439-030-9

AIChE